32ND
EDITION

A DESIGN COST DATA COMPANY
DATA YOU CAN TRUST

HOME BUILDER'S
2024 COSTBOOK

DATA YOU CAN TRUST

EDITOR-IN-CHIEF

William D. Mahoney, P.E.

TECHNICAL SERVICES

Joan Hamilton
Eric Mahoney, AIA
Ana Varela

GRAPHIC DESIGN

Robert O. Wright Jr.

BNi Publications, Inc.

VISTA
990 PARK CENTER DRIVE, SUITE E
VISTA, CA 92081

1-888-BNI-BOOK (1-888-264-2665)
www.bnibooks.com

ISBN 978-1-58855-248-8

Table of Contents

Preface

For over 75 years, BNi Building News has been dedicated to providing construction professionals with timely and reliable information. Based on this experience, our staff has researched and compiled thousands of up-to-the-minute costs for the **BNi Costbooks**. This book is an essential reference for contractors, engineers, architects, facility managers — any construction professional who must provide an estimate for any type of building project.

Whether working up a preliminary estimate or submitting a formal bid, the costs listed here can be quickly and easily tailored to your needs. All costs are based on prevailing labor rates. Overhead and profit should be included in all costs. Man-hours are also provided.

All data is categorized according to the CSI division format. This industry standard provides an all-inclusive checklist to ensure that no element of a project is overlooked. In addition, to make specific items even easier to locate, there is a complete alphabetical index.

The "Features of this Book" section presents a clear overview of the many features of this book. Included is an explanation of the data, sample page layout and discussion of how to best use the information in the book.

Of course, all buildings and construction projects are unique. The information provided in this book is based on averages from well-managed projects with good labor productivity under normal working conditions (eight hours a day). Other circumstances affecting costs, such as overtime, unusual working conditions, savings from buying bulk quantities for large projects, and unusual or hidden costs, must be factored in as they arise.

The data provided in this book is for estimating purposes only. Check all applicable federal, state and local codes and regulations for local requirements.

Format

All data is categorized according to the *CSI MASTERFORMAT*. This industry standard provides an all-inclusive checklist to ensure that no element of a project is overlooked.

DIVISION 00 ...PROCUREMENT & CONTRACTING REQUIREMENTS

00 10 00 SOLICITATION
00 20 00 INSTRUCTIONS FOR PROCUREMENT
00 30 00 AVAILABLE INFORMATION
00 40 00 PROCUREMENT FORMS AND SUPPLEMENTS
00 50 00 CONTRACTING FORMS AND SUPPLEMENTS
00 60 00 PROJECT FORMS
00 70 00 CONDITIONS OF THE CONTRACT
00 80 00 Reserved
00 90 00 REVISIONS, CLARIFICATIONS, AND MODIFICATIONS

DIVISION 01 .... GENERAL REQUIREMENTS

01 10 00 SUMMARY
01 20 00 PRICE AND PAYMENT PROCEDURES
01 30 00 ADMINISTRATIVE REQUIREMENTS
01 40 00 QUALITY REQUIREMENTS
01 50 00 TEMPORARY FACILITIES AND CONTROLS
01 60 00 PRODUCT REQUIREMENTS
01 70 00 EXECUTION AND CLOSEOUT REQUIREMENTS
01 80 00 PERFORMANCE REQUIREMENTS
01 90 00 LIFE CYCLE ACTIVITIES

DIVISION 02 .... EXISTING CONDITIONS

02 30 00 SUBSURFACE INVESTIGATION
02 40 00 DEMOLITION AND STRUCTURE MOVING
02 50 00 SITE REMEDIATION
02 60 00 CONTAMINATED SITE MATERIAL REMOVAL
02 70 00 WATER REMEDIATION
02 80 00 FACILITY REMEDIATION

DIVISION 03 .... CONCRETE

03 10 00 CONCRETE FORMING AND ACCESSORIES
03 20 00 CONCRETE REINFORCING
03 30 00 CAST-IN-PLACE CONCRETE
03 40 00 PRECAST CONCRETE
03 50 00 CAST DECKS AND UNDERLAYMENT
03 60 00 GROUTING
03 70 00 MASS CONCRETE
03 80 00 CONCRETE CUTTING AND BORING

DIVISION 04 .... MASONRY

04 20 00 UNIT MASONRY
04 30 00 Reserved
04 40 00 STONE ASSEMBLIES
04 50 00 REFRACTORY MASONRY
04 60 00 CORROSION-RESISTANT MASONRY
04 70 00 MANUFACTURED MASONRY
04 80 00 Reserved
04 90 00 Reserved

DIVISION 05...METALS

05 10 00 STRUCTURAL METAL FRAMING
05 12 00 STRUCTURAL STEEL FRAMING
05 20 00 METAL JOISTS
05 30 00 METAL DECKING
05 40 00 COLD-FORMED METAL FRAMING
05 50 00 METAL FABRICATIONS
05 60 00 Reserved
05 70 00 DECORATIVE METAL
05 80 00 Reserved
05 90 00 Reserved

DIVISION 06............................WOOD, PLASTICS, AND COMPOSITES

06 10 00 ROUGH CARPENTRY
06 30 00 Reserved
06 40 00 ARCHITECTURAL WOODWORK
06 50 00 STRUCTURAL PLASTICS
06 60 00 PLASTIC FABRICATIONS
06 70 00 STRUCTURAL COMPOSITES
06 80 00 COMPOSITE FABRICATIONS
06 90 00 Reserved

DIVISION 07.......................THERMAL AND MOISTURE PROTECTION

07 10 00 DAMPPROOFING AND WATERPROOFING
07 30 00 STEEP SLOPE ROOFING
07 40 00 ROOFING AND SIDING PANELS
07 50 00 MEMBRANE ROOFING
07 60 00 FLASHING AND SHEET METAL
07 70 00 ROOF AND WALL SPECIALTIES AND ACCESSORIES
07 80 00 FIRE AND SMOKE PROTECTION
07 90 00 JOINT PROTECTION

DIVISION 08..OPENINGS

08 10 00 DOORS AND FRAMES
08 20 00 Reserved
08 30 00 SPECIALTY DOORS AND FRAMES
08 40 00 ENTRANCES, STOREFRONTS, AND CURTAIN WALLS
08 50 00 WINDOWS
08 60 00 ROOF WINDOWS AND SKYLIGHTS
08 70 00 HARDWARE
08 80 00 GLAZING
08 90 00 LOUVERS AND VENTS

DIVISION 09...FINISHES

09 20 00 PLASTER AND GYPSUM BOARD
09 30 00 TILING
09 40 00 Reserved
09 50 00 CEILINGS
09 60 00 FLOORING
09 70 00 WALL FINISHES
09 80 00 ACOUSTIC TREATMENT
09 90 00 PAINTING AND COATING

Format *(Continued)*

Features of this Book

Sample pages with graphic explanations are included before the Costbook pages. These explanations, along with the discussions below, will provide a good understanding of what is included in this book and how it can best be used in construction estimating.

Material Costs

The material costs used in this book represent national averages for prices that a contractor would expect to pay plus an allowance for freight (if applicable) and handling and storage. These costs reflect neither the lowest or highest prices, but rather a typical average cost over time. Periodic fluctuations in availability and in certain commodities can significantly affect local material pricing. In the final estimating and bidding stages of a project when the highest degree of accuracy is required, it is best to check local, current prices.

Labor Costs

Labor costs include the basic wage, plus commonly applicable taxes, insurance and markups for overhead and profit. The labor rates used here to develop the costs are typical average prevailing wage rates. Rates for different trades are used where appropriate for each type of work.

Fixed government rates and average allowances for taxes and insurance are included in the labor costs. These include employer-paid Social Security/Medicare taxes (FICA), Worker's Compensation insurance, state and federal unemployment taxes, and business insurance.

Please note, however, most of these items vary significantly from state to state and within states. For more specific data, local agencies and sources should be consulted.

Man-Hours

These productivities represent typical installation labor for thousands of construction items. The data takes into account all activities involved in normal construction under commonly experienced working conditions such as site movement, material handling, start-up, etc.

Equipment Costs

Costs for various types and pieces of equipment are included in Division 1 - General Requirements and can be included in an estimate when required either as a total "Equipment" category or with specific appropriate trades. Costs for equipment are included when appropriate in the installation costs in the Costbook pages.

Overhead and Profit

Included in the labor costs are allowances for overhead and profit for the contractor/employer whose workers are performing the specific tasks. No cost allowances or fees are included for management of subcontractors by the general contractor or construction manager. These costs, where appropriate, must be added to the costs as listed in the book.

The allowance for overhead is included to account for office overhead, the contractors' typical costs of doing business. These costs normally include in-house office staff salaries and benefits, office rent and operating expenses, professional fees, vehicle costs and other operating costs which are not directly applicable to specific jobs. It should be noted for this book that office overhead as included should be distinguished from project overhead, the General Requirements (Division 1) which are specific to particular projects. Project overhead should be included on an item by item basis for each job.

Depending on the trade, an allowance of 10-15 percent is incorporated into the labor/installation costs to account for typical profit of the installing contractor. See Division 1, General Requirements, for a more detailed review of typical profit allowances.

Features of this Book *(Continued)*

Adjustments to Costs

The costs as presented in this book attempt to represent national averages. Costs, however, vary among regions, states and even between adjacent localities.

In order to more closely approximate the probable costs for specific locations throughout the U.S., a table of Geographic Multipliers is provided. These adjustment factors are used to modify costs obtained from this book to help account for regional variations of construction costs. Whenever local current costs are known, whether material or equipment prices or labor rates, they should be used if more accuracy is required.

Editor's Note: This **Costbook** is intended to provide accurate, reliable, average costs and typical productivities for thousands of common construction components. The data is developed and compiled from various industry sources, including government, manufacturers, suppliers and working professionals. The intent of the information is to provide assistance and guidelines to construction professionals in estimating. The user should be aware that local conditions, material and labor availability and cost variations, economic considerations, weather, local codes and regulations, etc., all affect the actual cost of construction. These and other such factors must be considered and incorporated into any and all construction estimates.

Sample Costbook Page

In order to best use the information in this book, please review this sample page and read the "Features of this Book" section.

Division

Broadscope Category (First 2 Digits)

Mediumscope Category (5 Digits)

Detailed Descriptions
Complete descriptions of items may include information listed above a particular line. Review of the whole category is recommended for a complete description.

Labor Cost
Labor cost represents U.S. prevailing wages plus applicable fringes.

Material Cost
Material cost represents average contractor prices plus an allowance for freight, handling and storage.

Equipment Cost
This cost includes equipment costs only, the wages for the crew operating the equipment are included in the Labor column.

Unit of Measurement
Each item (and cost) is defined in terms of the common estimating unit. All costs are listed in dollars per unit.

Total Cost
The total cost is the sum of material and installation costs. This total represents typical contractors' costs including overhead and profit, but does not include markups for the general contractor or construction management fees.

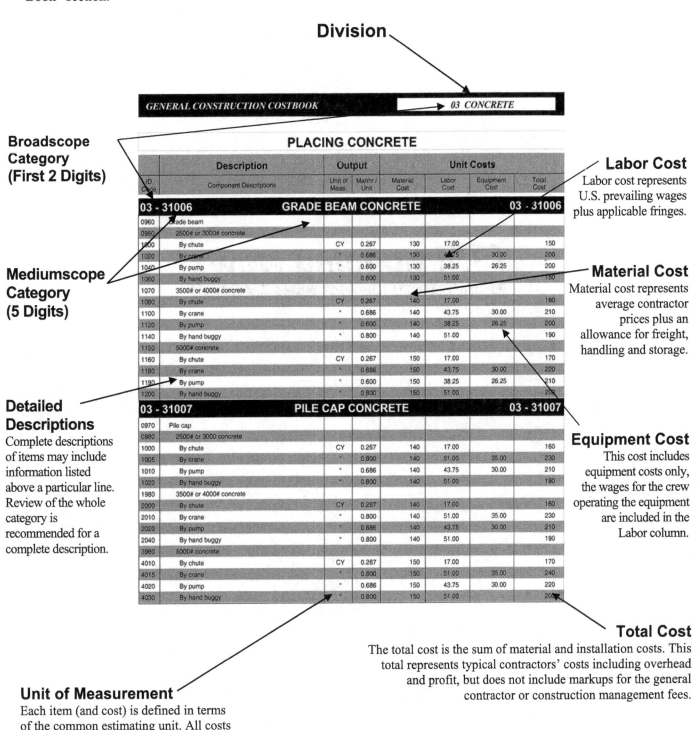

GENERAL CONSTRUCTION COSTBOOK — 03 CONCRETE

PLACING CONCRETE

ID Code	Description / Component Descriptions	Output Unit of Meas.	Output Manhr / Unit	Unit Costs Material Cost	Unit Costs Labor Cost	Unit Costs Equipment Cost	Unit Costs Total Cost
03 - 31006	GRADE BEAM CONCRETE						03 - 31006
0960	Grade beam						
0980	2500# or 3000# concrete						
1000	By chute	CY	0.267	130	17.00		150
1020	By crane	"	0.686	130	43.75	30.00	200
1040	By pump	"	0.600	130	38.25	26.25	200
1060	By hand buggy	"	0.800	130	51.00		180
1070	3500# or 4000# concrete						
1080	By chute	CY	0.267	140	17.00		160
1100	By crane	"	0.686	140	43.75	30.00	210
1120	By pump	"	0.600	140	38.25	26.25	200
1140	By hand buggy	"	0.800	140	51.00		190
1150	5000# concrete						
1160	By chute	CY	0.267	150	17.00		170
1180	By crane	"	0.686	150	43.75	30.00	220
1190	By pump	"	0.600	150	38.25	26.25	210
1200	By hand buggy	"	0.800	150	51.00		200
03 - 31007	PILE CAP CONCRETE						03 - 31007
0970	Pile cap						
0980	2500# or 3000 concrete						
1000	By chute	CY	0.267	140	17.00		160
1005	By crane	"	0.800	140	51.00	35.00	230
1010	By pump	"	0.686	140	43.75	30.00	210
1020	By hand buggy	"	0.800	140	51.00		190
1980	3500# or 4000# concrete						
2000	By chute	CY	0.267	140	17.00		160
2010	By crane	"	0.800	140	51.00	35.00	230
2020	By pump	"	0.686	140	43.75	30.00	210
2040	By hand buggy	"	0.800	140	51.00		190
3980	5000# concrete						
4010	By chute	CY	0.267	150	17.00		170
4015	By crane	"	0.800	150	51.00	35.00	240
4020	By pump	"	0.686	150	43.75	30.00	220
4030	By hand buggy	"	0.800	150	51.00		200

DIVISION 01
GENERAL

REQUIREMENTS

ID Code	Description — Component Descriptions	Output — Unit of Meas.	Output — Manhr / Unit	Unit Costs — Material Cost	Unit Costs — Labor Cost	Unit Costs — Equipment Cost	Total Cost
01 - 21001	**ALLOWANCES**						**01 - 21001**
0090	Overhead						
1000	$20,000 project						
1020	Minimum	PCT					15.00
1040	Average	"					20.00
1060	Maximum	"					40.00
1080	$100,000 project						
1100	Minimum	PCT					12.00
1120	Average	"					15.00
1140	Maximum	"					25.00
1160	$500,000 project						
1170	Minimum	PCT					10.00
1180	Average	"					12.00
1200	Maximum	"					20.00
1480	Profit						
1500	$20,000 project						
1520	Minimum	PCT					10.00
1540	Average	"					15.00
1560	Maximum	"					25.00
1580	$100,000 project						
1600	Minimum	PCT					10.00
1620	Average	"					12.00
1640	Maximum	"					20.00
1660	$500,000 project						
1680	Minimum	PCT					5.00
1700	Average	"					10.00
1720	Maximum	"					15.00
2000	Professional fees						
2100	Architectural						
2120	$100,000 project						
2140	Minimum	PCT					5.00
2160	Average	"					10.00
2180	Maximum	"					20.00
2200	$500,000 project						
2220	Minimum	PCT					5.00
2240	Average	"					8.00
2260	Maximum	"					12.00
2360	Structural engineering						
2380	Minimum	PCT					2.00

REQUIREMENTS

ID Code	Component Descriptions	Unit of Meas.	Manhr / Unit	Material Cost	Labor Cost	Equipment Cost	Total Cost
01 - 21001	**ALLOWANCES, Cont'd...**						**01 - 21001**
2400	Average	PCT					3.00
2420	Maximum	"					5.00
2440	Mechanical engineering						
2460	Minimum	PCT					4.00
2480	Average	"					5.00
2500	Maximum	"					15.00
4080	Taxes						
5000	Sales tax						
5020	Minimum	PCT					4.00
5040	Average	"					5.00
5060	Maximum	"					10.00
5080	Unemployment						
5100	Minimum	PCT					3.00
5120	Average	"					6.50
5140	Maximum	"					8.00
5200	Social security (FICA)	"					7.85
01 - 31130	**FIELD STAFF**						**01 - 31130**
1000	Superintendent						
1020	Minimum	YEAR					110,437
1040	Average	"					138,066
1060	Maximum	"					165,846
1160	Foreman						
1180	Minimum	YEAR					73,404
1200	Average	"					117,391
1220	Maximum	"					137,442
1240	Bookkeeper/timekeeper						
1260	Minimum	YEAR					42,460
1280	Average	"					55,448
1300	Maximum	"					71,739
1320	Watchman						
1340	Minimum	YEAR					31,637
1360	Average	"					42,322
1380	Maximum	"					53,422

REQUIREMENTS

ID Code	Component Descriptions	Unit of Meas.	Manhr / Unit	Material Cost	Labor Cost	Equipment Cost	Total Cost
	Description	**Output**		**Unit Costs**			
01 - 32230	**SURVEYING**						**01 - 32230**
0080	Surveying						
1000	Small crew	DAY					1,150
1020	Average crew	"					1,720
1040	Large crew	"					2,270
2000	Lot lines and boundaries						
2020	Minimum	ACRE					820
2040	Average	"					1,690
2060	Maximum	"					2,760
01 - 32330	**JOB REQUIREMENTS**						**01 - 32330**
1000	Job photographs, small jobs						
1020	Minimum	EA					160
1040	Average	"					250
1060	Maximum	"					580
1080	Large projects						
1100	Minimum	EA					820
1120	Average	"					1,230
1140	Maximum	"					4,110
01 - 45230	**TESTING**						**01 - 45230**
1080	Testing concrete, per test						
1100	Minimum	EA					26.75
1120	Average	"					44.50
1140	Maximum	"					89.00
01 - 54001	**CONSTRUCTION AIDS**						**01 - 54001**
1000	Scaffolding/staging, rent per month						
1020	Measured by lineal feet of base						
1040	10' high	LF					15.75
1060	20' high	"					28.50
1080	30' high	"					40.00
1140	Measured by square foot of surface						
1160	Minimum	SF					0.69
1180	Average	"					1.20
1200	Maximum	"					2.15
1300	Tarpaulins, fabric, per job						
1320	Minimum	SF					0.32
1340	Average	"					0.55
1360	Maximum	"					1.43

REQUIREMENTS

ID Code	Component Descriptions	Unit of Meas.	Manhr / Unit	Material Cost	Labor Cost	Equipment Cost	Total Cost
	Description	**Output**		**Unit Costs**			
01 - 54008	**MOBILIZATION**						**01 - 54008**
1000	Equipment mobilization						
1020	Bulldozer						
1040	Minimum	EA					260
1060	Average	"					540
1080	Maximum	"					910
1100	Backhoe/front-end loader						
1120	Minimum	EA					160
1140	Average	"					270
1160	Maximum	"					590
1260	Truck crane						
1280	Minimum	EA					650
1300	Average	"					1,010
1320	Maximum	"					1,730
01 - 54009	**EQUIPMENT**						**01 - 54009**
0080	Air compressor						
1000	60 cfm						
1020	By day	EA					120
1030	By week	"					350
1040	By month	"					1,050
1100	300 cfm						
1120	By day	EA					240
1130	By week	"					760
1140	By month	"					2,310
1300	Air tools, per compressor, per day						
1310	Minimum	EA					48.00
1320	Average	"					60.00
1330	Maximum	"					84.00
1400	Generators, 5 kw						
1410	By day	EA					120
1420	By week	"					360
1430	By month	"					1,100
1500	Heaters, salamander type, per week						
1510	Minimum	EA					140
1520	Average	"					200
1530	Maximum	"					430
1600	Pumps, submersible						
1605	50 gpm						

REQUIREMENTS

ID Code	Component Descriptions	Unit of Meas.	Manhr / Unit	Material Cost	Labor Cost	Equipment Cost	Total Cost
	Description	**Output**		**Unit Costs**			
01 - 54009	**EQUIPMENT, Cont'd...**						**01 - 54009**
1610	By day	EA					96.00
1620	By week	"					290
1630	By month	"					860
2000	Pickup truck						
2020	By day	EA					160
2030	By week	"					470
2040	By month	"					1,470
2080	Dump truck						
2100	6 c.y. truck						
2120	By day	EA					430
2130	By week	"					1,300
2140	By month	"					3,900
2160	10 c.y. truck						
2170	By day	EA					540
2180	By week	"					1,620
2190	By month	"					4,870
2300	16 c.y. truck						
2310	By day	EA					870
2320	By week	"					2,590
2340	By month	"					7,800
2400	Backhoe, track mounted						
2420	1/2 c.y. capacity						
2430	By day	EA					670
2440	By week	"					1,620
2450	By month	"					3,680
3000	Backhoe/loader, rubber tired						
3005	1/2 c.y. capacity						
3010	By day	EA					600
3020	By week	"					1,800
3030	By month	"					5,420
3035	3/4 c.y. capacity						
3040	By day	EA					720
3050	By week	"					2,160
3060	By month	"					6,500
3200	Bulldozer						
3205	75 hp						
3210	By day	EA					840
3220	By week	"					2,530

REQUIREMENTS

ID Code	Component Descriptions	Unit of Meas.	Manhr / Unit	Material Cost	Labor Cost	Equipment Cost	Total Cost
	Description	**Output**		**Unit Costs**			
01 - 54009	**EQUIPMENT, Cont'd...**						**01 - 54009**
3230	By month	EA					7,590
4000	Cranes, crawler type						
4005	15 ton capacity						
4010	By day	EA					1,080
4020	By week	"					3,250
4030	By month	"					9,750
4145	Truck mounted, hydraulic						
4150	15 ton capacity						
4160	By day	EA					1,020
4170	By week	"					3,070
4180	By month	"					8,860
5380	Loader, rubber tired						
5385	1 c.y. capacity						
5390	By day	EA					720
5400	By week	"					2,170
5410	By month	"					6,500
01 - 56230	**TEMPORARY FACILITIES**						**01 - 56230**
1000	Barricades, temporary						
1010	Highway						
1020	Concrete	LF	0.080	42.00	5.56		47.50
1040	Wood	"	0.032	9.61	2.22		11.75
1060	Steel	"	0.027	8.74	1.85		10.50
1090	Pedestrian barricades						
1100	Plywood	SF	0.027	8.74	1.85		10.50
1120	Chain link fence	"	0.027	6.99	1.85		8.84
1130	Trailers, general office type, per month						
2020	Minimum	EA					270
2040	Average	"					440
2060	Maximum	"					890
2070	Crew change trailers, per month						
2100	Minimum	EA					160
2120	Average	"					180
2140	Maximum	"					270

REQUIREMENTS

ID Code	Description		Output		Unit Costs			
	Component Descriptions		Unit of Meas.	Manhr / Unit	Material Cost	Labor Cost	Equipment Cost	Total Cost
01 - 58130			**SIGNS**					**01 - 58130**
0080	Construction signs, temporary							
1000	Signs, 2' x 4'							
1020	Minimum		EA					44.50
1040	Average		"					110
1060	Maximum		"					380
1080	Signs, 4' x 8'							
1100	Minimum		EA					94.00
1120	Average		"					240
1140	Maximum		"					1,050
1160	Signs, 8' x 8'							
1180	Minimum		EA					120
1200	Average		"					380
1220	Maximum		"					3,780

DIVISION 02
SITE CONSTRUCTION

SITE PREPARATION

ID Code	Component Descriptions	Unit of Meas.	Manhr / Unit	Material Cost	Labor Cost	Equipment Cost	Total Cost
	Description	**Output**		**Unit Costs**			

02 - 32130   SOIL BORING   02 - 32130

ID Code	Component Descriptions	Unit of Meas.	Manhr / Unit	Material Cost	Labor Cost	Equipment Cost	Total Cost
1000	Borings, uncased, stable earth						
1010	2-1/2" dia.	LF	0.300		20.75	19.75	40.50
1040	4" dia.	"	0.343		23.75	22.50	46.25
1500	Cased, including samples						
1520	2-1/2" dia.	LF	0.400		27.75	26.25	54.00
1540	4" dia.	"	0.686		47.50	45.00	92.00
2000	Drilling in rock						
2020	No sampling	LF	0.632		43.75	41.50	85.00
2040	With casing and sampling	"	0.800		55.00	53.00	110
3000	Test pits						
3020	Light soil	EA	4.000		280	260	540
3040	Heavy soil	"	6.000		410	390	810

SELECTIVE SITE DEMOLITION

02 - 41132   FENCE DEMOLITION   02 - 41132

ID Code	Component Descriptions	Unit of Meas.	Manhr / Unit	Material Cost	Labor Cost	Equipment Cost	Total Cost
0060	Remove fencing						
0080	Chain link, 8' high						
0100	For disposal	LF	0.040		2.78		2.78
0200	For reuse	"	0.100		6.96		6.96
0980	Wood						
1000	4' high	SF	0.027		1.85		1.85
1960	Masonry						
1980	8" thick						
2000	4' high	SF	0.080		5.56		5.56
2020	6' high	"	0.100		6.96		6.96

02 - 41133   CURB & GUTTER DEMOLITION   02 - 41133

ID Code	Component Descriptions	Unit of Meas.	Manhr / Unit	Material Cost	Labor Cost	Equipment Cost	Total Cost
0100	Removal, plain concrete curb	LF	0.060		4.14	3.93	8.08
0200	Plain concrete curb and 2' gutter	"	0.083		5.71	5.43	11.25

02 - 42132   GUARDRAIL DEMOLITION   02 - 42132

ID Code	Component Descriptions	Unit of Meas.	Manhr / Unit	Material Cost	Labor Cost	Equipment Cost	Total Cost
0080	Remove standard guardrail						
0085	Steel	LF	0.080		5.52	5.25	10.75
0090	Wood	"	0.062		4.25	4.03	8.29

SELECTIVE SITE DEMOLITION

ID Code	Component Descriptions	Unit of Meas.	Manhr / Unit	Material Cost	Labor Cost	Equipment Cost	Total Cost
	Description	**Output**		**Unit Costs**			
02 - 43133	**HYDRANT DEMOLITION**						**02 - 43133**
5000	Remove and reset fire hydrant	EA	12.000		830	790	1,620
02 - 44134	**PAVEMENT AND SIDEWALK DEMOLITION**						**02 - 44134**
0300	Concrete pavement, 6" thick						
0400	No reinforcement	SY	0.160		11.00	10.50	21.50
0450	With wire mesh	"	0.240		16.50	15.75	32.25
0550	With rebars	"	0.300		20.75	19.75	40.50
1500	Sidewalk, 4" thick, with disposal	"	0.080		5.52	5.25	10.75
02 - 45134	**DRAINAGE PIPING DEMOLITION**						**02 - 45134**
0100	Remove drainage pipe, not including excavation						
0200	12" dia.	LF	0.100		6.91	6.56	13.50
0300	18" dia.	"	0.126		8.72	8.28	17.00
02 - 46134	**GAS PIPING DEMOLITION**						**02 - 46134**
0980	Remove welded steel pipe, not including excavation						
1000	4" dia.	LF	0.150		10.25	9.84	20.25
2000	5" dia.	"	0.240		16.50	15.75	32.25
02 - 47134	**SANITARY PIPING DEMOLITION**						**02 - 47134**
0980	Remove sewer pipe, not including excavation						
1000	4" dia.	LF	0.096		6.63	6.30	13.00
02 - 48134	**WATER PIPING DEMOLITION**						**02 - 48134**
0980	Remove water pipe, not including excavation						
1000	4" dia.	LF	0.109		7.53	7.15	14.75
02 - 49135	**SAW CUTTING PAVEMENT**						**02 - 49135**
0100	Pavement, bituminous						
0110	2" thick	LF	0.016		1.10	1.43	2.53
0120	3" thick	"	0.020		1.38	1.78	3.16
0200	Concrete pavement, with wire mesh						
0210	4" thick	LF	0.031		2.12	2.75	4.87
0212	5" thick	"	0.033		2.30	2.97	5.28
0300	Plain concrete, unreinforced						
0320	4" thick	LF	0.027		1.84	2.38	4.22
0340	5" thick	"	0.031		2.12	2.75	4.87

DEMOLITION

ID Code	Component Descriptions	Unit of Meas.	Manhr / Unit	Material Cost	Labor Cost	Equipment Cost	Total Cost
	Description	**Output**		**Unit Costs**			
02 - 51061	**COMPLETE BUILDING DEMOLITION**						**02 - 51061**
0200	Wood frame	CF	0.003		0.18	0.29	0.47
02 - 51190	**SELECTIVE BUILDING DEMOLITION**						**02 - 51190**
1000	Partition removal						
1100	Concrete block partitions						
1140	8" thick	SF	0.053		3.71		3.71
1200	Brick masonry partitions						
1220	4" thick	SF	0.040		2.78		2.78
1240	8" thick	"	0.050		3.48		3.48
1700	Stud partitions						
1720	Metal or wood, with drywall both sides	SF	0.040		2.78		2.78
2000	Door and frame removal						
2140	Wood in framed wall						
2160	2'6"x6'8"	EA	0.571		39.75		39.75
2180	3'x6'8"	"	0.667		46.50		46.50
2500	Ceiling removal						
2520	Acoustical tile ceiling						
2540	Adhesive fastened	SF	0.008		0.55		0.55
2560	Furred and glued	"	0.007		0.46		0.46
2580	Suspended grid	"	0.005		0.34		0.34
2600	Drywall ceiling						
2620	Furred and nailed	SF	0.009		0.61		0.61
2640	Nailed to framing	"	0.008		0.55		0.55
9300	Window removal						
9301	Metal windows, trim included						
9302	2'x3'	EA	0.800		56.00		56.00
9308	3'x4'	"	1.000		70.00		70.00
9316	4'x8'	"	2.000		140		140
9317	Wood windows, trim included						
9318	2'x3'	EA	0.444		31.00		31.00
9321	3'x4'	"	0.533		37.00		37.00
9327	6'x8'	"	0.800		56.00		56.00
9420	Concrete block walls, not including toothing						
9440	4" thick	SF	0.044		3.09		3.09
9450	6" thick	"	0.047		3.27		3.27
9460	8" thick	"	0.050		3.48		3.48
9500	Rubbish handling						
9519	Load in dumpster or truck						

DEMOLITION

ID Code	Component Descriptions	Unit of Meas.	Manhr / Unit	Material Cost	Labor Cost	Equipment Cost	Total Cost
	Description	**Output**		**Unit Costs**			
02 - 51190	**SELECTIVE BUILDING DEMOLITION, Cont'd...**						**02 - 51190**
9520	Minimum	CF	0.018		1.23		1.23
9540	Maximum	"	0.027		1.85		1.85
9600	Rubbish hauling						
9640	Hand loaded on trucks, 2 mile trip	CY	0.320		22.00	28.50	51.00
9660	Machine loaded on trucks, 2 mile trip	"	0.240		16.50	15.75	32.25

SITE REMEDIATION

ID Code	Component Descriptions	Unit of Meas.	Manhr / Unit	Material Cost	Labor Cost	Equipment Cost	Total Cost
02 - 65007	**SEPTIC TANK REMOVAL**						**02 - 65007**
0980	Remove septic tank						
1000	1000 gals	EA	2.000		140	130	270
1020	2000 gals	"	2.400		170	160	320

DIVISION 03
CONCRETE

FORMWORK

ID Code	Description	Output		Unit Costs			
	Component Descriptions	Unit of Meas.	Manhr / Unit	Material Cost	Labor Cost	Equipment Cost	Total Cost

03 - 11130 BEAM FORMWORK 03 - 11130

ID	Component	U/M	Manhr	Matl	Labor	Equip	Total
1000	Beam forms, job built						
1020	Beam bottoms						
1040	1 use	SF	0.133	5.82	11.75		17.50
1100	4 uses	"	0.118	2.17	10.50		12.75
1120	5 uses	"	0.114	1.98	10.25		12.25
2000	Beam sides						
2020	1 use	SF	0.089	4.16	7.89		12.00
2100	5 uses	"	0.073	1.76	6.45		8.21

03 - 11132 COLUMN FORMWORK 03 - 11132

ID	Component	U/M	Manhr	Matl	Labor	Equip	Total
1000	Column, square forms, job built						
1020	8" x 8" columns						
1040	1 use	SF	0.160	4.90	14.25		19.25
1120	5 uses	"	0.138	1.73	12.25		14.00
1200	12" x 12" columns						
1220	1 use	SF	0.145	4.46	13.00		17.50
1290	5 uses	"	0.127	1.45	11.25		12.75
2000	Round fiber forms, 1 use						
2040	10" dia.	LF	0.160	5.79	14.25		20.00
2060	12" dia.	"	0.163	7.13	14.50		21.75

03 - 11133 CURB FORMWORK 03 - 11133

ID	Component	U/M	Manhr	Matl	Labor	Equip	Total
0980	Curb forms						
0990	Straight, 6" high						
1000	1 use	LF	0.080	2.91	7.10		10.00
1080	5 uses	"	0.067	1.07	5.92		6.99
1090	Curved, 6" high						
2000	1 use	LF	0.100	3.16	8.88		12.00
2080	5 uses	"	0.082	1.28	7.24		8.52

03 - 11135 EQUIPMENT PAD FORMWORK 03 - 11135

ID	Component	U/M	Manhr	Matl	Labor	Equip	Total
1000	Equipment pad, job built						
1020	1 use	SF	0.100	5.08	8.88		14.00
1060	3 uses	"	0.089	2.45	7.89		10.25
1100	5 uses	"	0.080	1.51	7.10		8.61

FORMWORK

ID Code	Component Descriptions	Unit of Meas.	Manhr / Unit	Material Cost	Labor Cost	Equipment Cost	Total Cost
	Description	**Output**		**Unit Costs**			
03 - 11136	**FOOTING FORMWORK**						**03 - 11136**
2000	Wall footings, job built, continuous						
2040	1 use	SF	0.080	2.72	7.10		9.82
2060	3 uses	"	0.073	1.58	6.45		8.03
2090	5 uses	"	0.067	1.21	5.92		7.13
03 - 11137	**GRADE BEAM FORMWORK**						**03 - 11137**
1000	Grade beams, job built						
1020	1 use	SF	0.080	4.28	7.10		11.50
1060	3 uses	"	0.073	1.88	6.45		8.33
1100	5 uses	"	0.067	1.30	5.92		7.22
03 - 11138	**PILE CAP FORMWORK**						**03 - 11138**
1500	Pile cap forms, job built						
1510	Square						
1520	1 use	SF	0.100	4.86	8.88		13.75
1600	5 uses	"	0.080	1.63	7.10		8.73
03 - 11139	**SLAB / MAT FORMWORK**						**03 - 11139**
3000	Mat foundations, job built						
3020	1 use	SF	0.100	4.25	8.88		13.25
3060	3 uses	"	0.089	1.80	7.89		9.69
3100	5 uses	"	0.080	1.22	7.10		8.32
3980	Edge forms						
3990	6" high						
4000	1 use	LF	0.073	4.08	6.45		10.50
4002	3 uses	"	0.067	1.73	5.92		7.65
4004	5 uses	"	0.062	1.17	5.46		6.63
03 - 11141	**WALL FORMWORK**						**03 - 11141**
2980	Wall forms, exterior, job built						
3000	Up to 8' high wall						
3120	1 use	SF	0.080	4.57	7.10		11.75
3160	3 uses	"	0.073	2.22	6.45		8.67
3190	5 uses	"	0.067	1.68	5.92		7.60
4591	Retaining wall forms						
4592	1 use	SF	0.089	4.25	7.89		12.25
4594	3 uses	"	0.080	1.95	7.10		9.05
4596	5 uses	"	0.073	1.44	6.45		7.89
5000	Column pier and pilaster						

FORMWORK

ID Code	Description — Component Descriptions	Output — Unit of Meas.	Output — Manhr / Unit	Unit Costs — Material Cost	Unit Costs — Labor Cost	Unit Costs — Equipment Cost	Unit Costs — Total Cost
03 - 11141	**WALL FORMWORK, Cont'd...**						**03 - 11141**
5020	1 use	SF	0.160	4.81	14.25		19.00
5090	5 uses	"	0.114	2.16	10.25		12.50
03 - 11242	**MISCELLANEOUS FORMWORK**						**03 - 11242**
1200	Keyway forms (5 uses)						
1220	2 x 4	LF	0.040	0.37	3.55		3.92
1240	2 x 6	"	0.044	0.54	3.94		4.48
1500	Bulkheads						
1510	Walls, with keyways						
1520	3 piece	LF	0.080	7.08	7.10		14.25
1600	Ground slab, with keyway						
1620	2 piece	LF	0.057	6.64	5.07		11.75
1640	3 piece	"	0.062	8.10	5.46		13.50
2000	Chamfer strips						
2020	Wood						
2040	1/2" wide	LF	0.018	0.29	1.57		1.86
2060	3/4" wide	"	0.018	0.37	1.57		1.94
2070	1" wide	"	0.018	0.50	1.57		2.07
2100	PVC						
2120	1/2" wide	LF	0.018	1.29	1.57		2.86
2140	3/4" wide	"	0.018	1.40	1.57		2.97
2160	1" wide	"	0.018	2.03	1.57		3.60

ACCESSORIES

ID Code	Description — Component Descriptions	Output — Unit of Meas.	Output — Manhr / Unit	Unit Costs — Material Cost	Unit Costs — Labor Cost	Unit Costs — Equipment Cost	Unit Costs — Total Cost
03 - 15001	**CONCRETE ACCESSORIES**						**03 - 15001**
1000	Expansion joint, poured						
1010	Asphalt						
1020	1/2" x 1"	LF	0.016	1.19	1.11		2.30
1040	1" x 2"	"	0.017	3.73	1.21		4.94
1300	Expansion joint, premolded, in slabs						
1310	Asphalt						
1320	1/2" x 6"	LF	0.020	1.00	1.39		2.39
1340	1" x 12"	"	0.027	1.67	1.85		3.52
1350	Cork						
1360	1/2" x 6"	LF	0.020	1.99	1.39		3.38
1380	1" x 12"	"	0.027	7.56	1.85		9.41
1390	Neoprene sponge						
1400	1/2" x 6"	LF	0.020	3.37	1.39		4.76

ACCESSORIES

ID Code	Component Descriptions	Unit of Meas.	Manhr / Unit	Material Cost	Labor Cost	Equipment Cost	Total Cost
		Output		**Unit Costs**			
03 - 15001	**CONCRETE ACCESSORIES, Cont'd...**						**03 - 15001**
1420	1" x 12"	LF	0.027	12.25	1.85		14.00
1430	Polyethylene foam						
1440	1/2" x 6"	LF	0.020	1.13	1.39		2.52
1460	1" x 12"	"	0.027	5.21	1.85		7.06
1560	Polyurethane foam						
1580	1/2" x 6"	LF	0.020	1.49	1.39		2.88
1600	1" x 12"	"	0.027	3.85	1.85		5.70
1610	Polyvinyl chloride foam						
1620	1/2" x 6"	LF	0.020	3.67	1.39		5.06
1640	1" x 12"	"	0.027	7.92	1.85		9.77
1650	Rubber, gray sponge						
1660	1/2" x 6"	LF	0.020	4.97	1.39		6.36
1680	1" x 12"	"	0.027	21.50	1.85		23.25
1700	Asphalt felt control joints or bond breaker, screed joints						
1780	4" slab	LF	0.016	1.41	1.11		2.52
1800	6" slab	"	0.018	1.76	1.23		2.99
1820	8" slab	"	0.020	2.30	1.39		3.69
2100	Waterstops						
2120	Polyvinyl chloride						
2125	Ribbed						
2130	3/16" thick x						
2140	4" wide	LF	0.040	1.98	2.78		4.76
2160	6" wide	"	0.044	2.42	3.09		5.51
2165	1/2" thick x						
2170	9" wide	LF	0.050	6.43	3.48		9.91
2178	Ribbed with center bulb						
2180	3/16" thick x 9" wide	LF	0.050	5.40	3.48		8.88
2200	3/8" thick x 9" wide	"	0.050	6.35	3.48		9.83
2240	Dumbbell type, 3/8" thick x 6" wide	"	0.044	6.43	3.09		9.52
2260	Plain, 3/8" thick x 9" wide	"	0.050	8.54	3.48		12.00
2280	Center bulb, 3/8" thick x 9" wide	"	0.050	10.25	3.48		13.75
2300	Rubber						
5060	Vapor barrier						
5090	4 mil polyethylene	SF	0.003	0.05	0.18		0.23
5094	6 mil polyethylene	"	0.003	0.08	0.18		0.26
5096	10 mil polyethylene	"	0.003	0.13	0.18		0.31
5098	15 mil polyethylene	"	0.003	0.14	0.18		0.32
5200	Gravel porous fill, under floor slabs, 3/4" stone	CY	1.333	30.75	93.00		120

ACCESSORIES

ID Code	Description Component Descriptions	Output Unit of Meas.	Manhr / Unit	Unit Costs Material Cost	Labor Cost	Equipment Cost	Total Cost
03 - 15001	**CONCRETE ACCESSORIES, Cont'd...**						**03 - 15001**
6000	Reinforcing accessories						
6010	Beam bolsters						
6020	1-1/2" high, plain	LF	0.008	0.59	0.72		1.31
6030	Galvanized	"	0.008	1.30	0.72		2.02
6035	3" high						
6040	Plain	LF	0.010	0.84	0.90		1.74
6050	Galvanized	"	0.010	2.08	0.90		2.98
6080	Slab bolsters						
6090	1" high						
6100	Plain	LF	0.004	0.63	0.36		0.99
6110	Galvanized	"	0.004	1.27	0.36		1.63
6115	2" high						
6120	Plain	LF	0.004	0.73	0.40		1.13
6130	Galvanized	"	0.004	1.49	0.40		1.89
6210	Chairs, high chairs						
6215	2" high						
6218	Plastic	EA	0.020	0.82	1.81		2.63
6220	Plain	"	0.020	1.70	1.81		3.51
6230	Galvanized	"	0.020	1.87	1.81		3.68
6255	4-1/2" high						
6258	Plastic	EA	0.023	1.65	2.07		3.72
6260	Plain	"	0.023	2.88	2.07		4.95
6270	Galvanized	"	0.023	4.88	2.07		6.95
6295	Continuous, high chair						
6299	3" high						
6300	Plain	LF	0.005	2.36	0.48		2.84
6310	Galvanized	"	0.005	2.92	0.48		3.40

REINFORCEMENT

ID Code	Description Component Descriptions	Output Unit of Meas.	Manhr / Unit	Material Cost	Labor Cost	Equipment Cost	Total Cost
03 - 21001	**BEAM REINFORCING**						**03 - 21001**
0980	Beam-girders						
1000	#3 - #4	TON	20.000	1,880	1,810		3,690
1010	#5 - #6	"	16.000	1,650	1,450		3,100

REINFORCEMENT

ID Code	Component Descriptions	Unit of Meas.	Manhr / Unit	Material Cost	Labor Cost	Equipment Cost	Total Cost
03 - 21003	**COLUMN REINFORCING**						**03 - 21003**
0980	Columns						
1000	#3 - #4	TON	22.857	1,880	2,070		3,950
1010	#5 - #6	"	17.778	1,650	1,610		3,260
03 - 21004	**ELEVATED SLAB REINFORCING**						**03 - 21004**
0980	Elevated slab						
1000	#3 - #4	TON	10.000	1,880	910		2,790
1020	#5 - #6	"	8.889	1,650	810		2,460
03 - 21005	**EQUIP. PAD REINFORCING**						**03 - 21005**
0980	Equipment pad						
1000	#3 - #4	TON	16.000	1,880	1,450		3,330
1020	#5 - #6	"	14.545	1,650	1,320		2,970
03 - 21006	**FOOTING REINFORCING**						**03 - 21006**
1000	Footings						
1020	#3 - #4	TON	13.333	1,880	1,210		3,090
1030	#5 - #6	"	11.429	1,650	1,040		2,690
1040	#7 - #8	"	10.000	1,610	910		2,520
4980	Straight dowels, 24" long						
5040	3/4" dia. (#6)	EA	0.019	5.24	1.68		6.92
5050	5/8" dia. (#5)	"	0.017	4.53	1.57		6.10
5060	1/2" dia. (#4)	"	0.016	3.41	1.45		4.86
03 - 21007	**FOUNDATION REINFORCING**						**03 - 21007**
0980	Foundations						
1000	#3 - #4	TON	13.333	1,880	1,210		3,090
1020	#5 - #6	"	11.429	1,650	1,040		2,690
1040	#7 - #8	"	10.000	1,610	910		2,520
03 - 21008	**GRADE BEAM REINFORCING**						**03 - 21008**
0980	Grade beams						
1000	#3 - #4	TON	12.308	1,880	1,120		3,000
1020	#5 - #6	"	10.667	1,650	970		2,620
1040	#7 - #8	"	9.412	1,610	850		2,460

REINFORCEMENT

ID Code	Component Descriptions	Unit of Meas.	Manhr / Unit	Material Cost	Labor Cost	Equipment Cost	Total Cost
	Description	**Output**		**Unit Costs**			
03 - 21009	**SLAB / MAT REINFORCING**						**03 - 21009**
0900	Bars, slabs						
1000	#3 - #4	TON	13.333	1,880	1,210		3,090
1020	#5 - #6	"	11.429	1,650	1,040		2,690
4990	Wire mesh, slabs						
5000	Galvanized						
5005	2x2						
5006	W.9xW.9	SF	0.005	2.29	0.48		2.77
5010	4x4						
5020	W1.4xW1.4	SF	0.005	0.54	0.48		1.02
5040	W2.0xW2.0	"	0.006	0.70	0.51		1.21
5060	W2.9xW2.9	"	0.006	0.98	0.55		1.53
5080	W4.0xW4.0	"	0.007	1.45	0.60		2.05
5090	6x6						
5100	W1.4xW1.4	SF	0.004	0.49	0.36		0.85
5120	W2.0xW2.0	"	0.004	0.70	0.40		1.10
5140	W2.9xW2.9	"	0.005	0.94	0.42		1.36
5150	W4.0xW4.0	"	0.005	1.01	0.48		1.49
03 - 21011	**WALL REINFORCING**						**03 - 21011**
0980	Walls						
1000	#3 - #4	TON	11.429	1,880	1,040		2,920
1020	#5 - #6	"	10.000	1,650	910		2,560
8980	Masonry wall (horizontal)						
9000	#3 - #4	TON	32.000	1,880	2,900		4,780
9020	#5 - #6	"	26.667	1,650	2,420		4,070
9030	Galvanized						
9040	#3 - #4	TON	32.000	3,210	2,900		6,110
9060	#5 - #6	"	26.667	3,030	2,420		5,450
9180	Masonry wall (vertical)						
9200	#3 - #4	TON	40.000	1,880	3,630		5,510
9220	#5 - #6	"	32.000	1,650	2,900		4,550
9230	Galvanized						
9240	#3 - #4	TON	40.000	3,210	3,630		6,840
9260	#5 - #6	"	32.000	3,030	2,900		5,930

REINFORCEMENT

ID Code	Description Component Descriptions	Output Unit of Meas.	Output Manhr / Unit	Unit Costs Material Cost	Unit Costs Labor Cost	Unit Costs Equipment Cost	Unit Costs Total Cost
03 - 21016	**PILE CAP REINFORCING**						**03 - 21016**
0980	Pile caps						
1000	#3 - #4	TON	20.000	1,880	1,810		3,690
1020	#5 - #6	"	17.778	1,650	1,610		3,260
1040	#7 - #8	"	16.000	1,610	1,450		3,060

CAST-IN-PLACE CONCRETE

ID Code	Description Component Descriptions	Output Unit of Meas.	Output Manhr / Unit	Unit Costs Material Cost	Unit Costs Labor Cost	Unit Costs Equipment Cost	Unit Costs Total Cost
03 - 30531	**CONCRETE ADMIXTURES**						**03 - 30531**
1000	Concrete admixtures						
1020	Water reducing admixture	GAL					19.25
1040	Set retarder	"					16.00
1060	Air entraining agent	"					18.25

PLACING CONCRETE

ID Code	Description Component Descriptions	Output Unit of Meas.	Output Manhr / Unit	Unit Costs Material Cost	Unit Costs Labor Cost	Unit Costs Equipment Cost	Unit Costs Total Cost
03 - 31001	**BEAM CONCRETE**						**03 - 31001**
0960	Beams and girders						
0980	2500# or 3000# concrete						
1000	By crane	CY	0.960	170	67.00	50.00	290
1010	By pump	"	0.873	170	61.00	45.50	280
1020	By hand buggy	"	0.800	170	56.00		230
2480	3500# or 4000# concrete						
2500	By crane	CY	0.960	180	67.00	50.00	300
2520	By pump	"	0.873	180	61.00	45.50	290
2540	By hand buggy	"	0.800	180	56.00		240
03 - 31002	**COLUMN CONCRETE**						**03 - 31002**
0980	Columns						
0990	2500# or 3000# concrete						
1000	By crane	CY	0.873	170	61.00	45.50	280
1010	By pump	"	0.800	170	56.00	41.75	270
1980	3500# or 4000# concrete						
2000	By crane	CY	0.873	180	61.00	45.50	290
2020	By pump	"	0.800	180	56.00	41.75	280

PLACING CONCRETE

ID Code	Component Descriptions	Unit of Meas.	Manhr / Unit	Material Cost	Labor Cost	Equipment Cost	Total Cost
	Description	**Output**		**Unit Costs**			

03 - 31003 — ELEVATED SLAB CONCRETE — 03 - 31003

ID Code	Component Descriptions	Unit of Meas.	Manhr / Unit	Material Cost	Labor Cost	Equipment Cost	Total Cost
0980	Elevated slab						
0990	2500# or 3000# concrete						
1000	By crane	CY	0.480	170	33.50	25.00	230
1010	By pump	"	0.369	170	25.75	19.25	210
1020	By hand buggy	"	0.800	170	56.00		230

03 - 31004 — EQUIPMENT PAD CONCRETE — 03 - 31004

ID Code	Component Descriptions	Unit of Meas.	Manhr / Unit	Material Cost	Labor Cost	Equipment Cost	Total Cost
0960	Equipment pad						
0980	2500# or 3000# concrete						
1000	By chute	CY	0.267	170	18.50		190
1020	By pump	"	0.686	170	47.75	35.75	250
1040	By crane	"	0.800	170	56.00	41.75	270
1050	3500# or 4000# concrete						
1060	By chute	CY	0.267	180	18.50		200
1080	By pump	"	0.686	180	47.75	35.75	260

03 - 31005 — FOOTING CONCRETE — 03 - 31005

ID Code	Component Descriptions	Unit of Meas.	Manhr / Unit	Material Cost	Labor Cost	Equipment Cost	Total Cost
0980	Continuous footing						
0990	2500# or 3000# concrete						
1000	By chute	CY	0.267	170	18.50		190
1010	By pump	"	0.600	170	41.75	31.25	240
1020	By crane	"	0.686	170	47.75	35.75	250
4980	Spread footing						
5000	2500# or 3000# concrete						
5020	By chute	CY	0.267	170	18.50		190
5040	By pump	"	0.640	170	44.50	33.25	250
5060	By crane	"	0.738	170	51.00	38.50	260

03 - 31006 — GRADE BEAM CONCRETE — 03 - 31006

ID Code	Component Descriptions	Unit of Meas.	Manhr / Unit	Material Cost	Labor Cost	Equipment Cost	Total Cost
0960	Grade beam						
0980	2500# or 3000# concrete						
1000	By chute	CY	0.267	170	18.50		190
1020	By crane	"	0.686	170	47.75	35.75	250
1040	By pump	"	0.600	170	41.75	31.25	240
1060	By hand buggy	"	0.800	170	56.00		230
1070	3500# or 4000# concrete						
1080	By chute	CY	0.267	180	18.50		200
1100	By crane	"	0.686	180	47.75	35.75	260
1120	By pump	"	0.600	180	41.75	31.25	250

PLACING CONCRETE

ID Code	Description		Output		Unit Costs			
	Component Descriptions		Unit of Meas.	Manhr / Unit	Material Cost	Labor Cost	Equipment Cost	Total Cost
03 - 31006	**GRADE BEAM CONCRETE, Cont'd...**							**03 - 31006**
1140	By hand buggy		CY	0.800	180	56.00		240
03 - 31007	**PILE CAP CONCRETE**							**03 - 31007**
0970	Pile cap							
0980	2500# or 3000 concrete							
1000	By chute		CY	0.267	170	18.50		190
1005	By crane		"	0.800	170	56.00	41.75	270
1010	By pump		"	0.686	170	47.75	35.75	250
1020	By hand buggy		"	0.800	170	56.00		230
03 - 31008	**SLAB / MAT CONCRETE**							**03 - 31008**
0960	Slab on grade							
0980	2500# or 3000# concrete							
1000	By chute		CY	0.200	170	14.00		180
1010	By crane		"	0.400	170	27.75	20.75	220
1020	By pump		"	0.343	170	23.75	17.75	210
1030	By hand buggy		"	0.533	170	37.00		210
03 - 31009	**WALL CONCRETE**							**03 - 31009**
0940	Walls							
0960	2500# or 3000# concrete							
0980	To 4'							
1000	By chute		CY	0.229	170	16.00		190
1005	By crane		"	0.800	170	56.00	41.75	270
1010	By pump		"	0.738	170	51.00	38.50	260
1020	To 8'							
1030	By crane		CY	0.873	170	61.00	45.50	280
1040	By pump		"	0.800	170	56.00	41.75	270
8480	Filled block (CMU)							
8490	3000# concrete, by pump							
8500	4" wide		SF	0.034	2.08	2.38	1.78	6.25
8510	6" wide		"	0.040	3.10	2.78	2.08	7.96
8520	8" wide		"	0.048	4.17	3.34	2.50	10.00

CONCRETE FINISHING

ID Code	Component Descriptions	Unit of Meas.	Manhr / Unit	Material Cost	Labor Cost	Equipment Cost	Total Cost
	Description	**Output**		**Unit Costs**			
03 - 35001	**CONCRETE FINISHES**						**03 - 35001**
0980	Floor finishes						
1000	Broom	SF	0.011		0.79		0.79
1020	Screed	"	0.010		0.69		0.69
1040	Darby	"	0.010		0.69		0.69
1060	Steel float	"	0.013		0.92		0.92
4000	Wall finishes						
4020	Burlap rub, with cement paste	SF	0.013	0.14	0.92		1.06
03 - 37190	**PNEUMATIC CONCRETE**						**03 - 37190**
0100	Pneumatic applied concrete (gunite)						
1035	2" thick	SF	0.030	6.55	2.07	1.96	10.50
1040	3" thick	"	0.040	8.04	2.76	2.62	13.50
1060	4" thick	"	0.048	9.81	3.31	3.15	16.25
1980	Finish surface						
2000	Minimum	SF	0.040		3.55		3.55
2020	Maximum	"	0.080		7.10		7.10
03 - 39001	**CURING CONCRETE**						**03 - 39001**
1000	Sprayed membrane						
1010	Slabs	SF	0.002	0.11	0.11		0.22
1020	Walls	"	0.002	0.16	0.13		0.29
1025	Curing paper						
1030	Slabs	SF	0.002	0.16	0.14		0.30
2000	Walls	"	0.002	0.16	0.16		0.32
2010	Burlap						
2020	7.5 oz.	SF	0.003	0.14	0.18		0.32
2500	12 oz.	"	0.003	0.19	0.19		0.38
03 - 48001	**PRECAST SPECIALTIES**						**03 - 48001**
0980	Precast concrete, coping, 4' to 8' long						
1000	12" wide	LF	0.060	11.50	4.14	3.93	19.50
1010	10" wide	"	0.069	10.25	4.73	4.50	19.50
1520	Splash block, 30"x12"x4"	EA	0.400	17.50	27.75	26.25	72.00
2000	Stair unit, per riser	"	0.400	110	27.75	26.25	160
4000	Sun screen and trellis, 8' long, 12" high						
4020	4" thick blades	EA	0.300	120	20.75	19.75	160

GROUT

ID Code	Description		Output		Unit Costs			
	Component Descriptions		Unit of Meas.	Manhr / Unit	Material Cost	Labor Cost	Equipment Cost	Total Cost

03 - 61001 — GROUTING — 03 - 61001

ID Code	Component Descriptions	Unit of Meas.	Manhr / Unit	Material Cost	Labor Cost	Equipment Cost	Total Cost
1000	Grouting for bases						
2000	Non-metallic grout						
2020	1" deep	SF	0.160	6.68	11.25	4.20	22.00
2040	2" deep	"	0.178	12.75	12.25	4.66	29.75
3020	Portland cement grout (1 cement to 3 sand)						
3040	1/2" joint thickness						
3080	6" wide joints	LF	0.027	0.22	1.85	0.70	2.77
3100	8" wide joints	"	0.032	0.25	2.22	0.84	3.31
3200	1" joint thickness						
3220	4" wide joints	LF	0.025	0.25	1.74	0.65	2.64
3240	6" wide joints	"	0.028	0.41	1.92	0.72	3.05

CONCRETE BORING

03 - 82131 — CORE DRILLING — 03 - 82131

ID Code	Component Descriptions	Unit of Meas.	Manhr / Unit	Material Cost	Labor Cost	Equipment Cost	Total Cost
0100	Concrete						
0110	6" thick						
0120	3" dia.	EA	0.571		39.75	15.00	55.00
0300	8" thick						
0320	3" dia.	EA	0.800		56.00	21.00	77.00

DIVISION 04
MASONRY

MASONRY RESTORATION

ID Code	Component Descriptions	Unit of Meas.	Manhr / Unit	Material Cost	Labor Cost	Equipment Cost	Total Cost
	Description	**Output**		**Unit Costs**			

04 - 01201   RESTORATION AND CLEANING   04 - 01201

ID Code	Component Descriptions	Unit of Meas.	Manhr / Unit	Material Cost	Labor Cost	Equipment Cost	Total Cost
1080	Masonry cleaning						
1090	Washing brick						
1120	Smooth surface	SF	0.013	0.21	1.13		1.34
1130	Rough surface	"	0.018	0.29	1.50		1.79
1140	Steam clean masonry						
1150	Smooth face						
1220	Minimum	SF	0.010	0.56	0.69	0.26	1.51
1240	Maximum	"	0.015	0.90	1.01	0.38	2.29
1250	Rough face						
1260	Minimum	SF	0.013	0.82	0.92	0.35	2.09
1270	Maximum	"	0.020	1.21	1.39	0.52	3.12

MORTAR, GROUT AND ACCESSORIES

04 - 05161   MASONRY GROUT   04 - 05161

ID Code	Component Descriptions	Unit of Meas.	Manhr / Unit	Material Cost	Labor Cost	Equipment Cost	Total Cost
0100	Grout, non-shrink, non-metallic, trowelable	CF	0.016	5.09	1.10	1.05	7.24
2110	Grout door frame, hollow metal						
2120	Single	EA	0.600	12.75	41.50	39.25	94.00
2140	Double	"	0.632	17.75	43.75	41.50	100
2980	Grout-filled concrete block (CMU)						
3000	4" wide	SF	0.020	0.33	1.38	1.31	3.02
3020	6" wide	"	0.022	0.89	1.50	1.43	3.82
3040	8" wide	"	0.024	1.31	1.65	1.57	4.54
3060	12" wide	"	0.025	2.15	1.74	1.65	5.55
3070	Grout-filled individual CMU cells						
3090	4" wide	LF	0.012	0.28	0.82	0.78	1.89
3100	6" wide	"	0.012	0.38	0.82	0.78	1.99
3120	8" wide	"	0.012	0.50	0.82	0.78	2.11
3140	10" wide	"	0.014	0.63	0.94	0.90	2.47
3160	12" wide	"	0.014	0.77	0.94	0.90	2.61
4000	Bond beams or lintels, 8" deep						
4010	6" thick	LF	0.022	0.77	1.51	1.13	3.42
4020	8" thick	"	0.024	1.01	1.67	1.25	3.93
4040	10" thick	"	0.027	1.28	1.85	1.38	4.52
4060	12" thick	"	0.030	1.52	2.08	1.56	5.17
5000	Cavity walls						
5020	2" thick	SF	0.032	0.84	2.22	1.66	4.73
5040	3" thick	"	0.032	1.28	2.22	1.66	5.17

MORTAR, GROUT AND ACCESSORIES

ID Code	Component Descriptions	Unit of Meas.	Manhr / Unit	Material Cost	Labor Cost	Equipment Cost	Total Cost
04 - 05161	**MASONRY GROUT, Cont'd...**						**04 - 05161**
5060	4" thick	SF	0.034	1.69	2.38	1.78	5.86
5080	6" thick	"	0.040	2.54	2.78	2.08	7.40
04 - 05231	**MASONRY ACCESSORIES**						**04 - 05231**
0200	Foundation vents	EA	0.320	32.25	27.00		59.00
1010	Bar reinforcing						
1015	Horizontal						
1020	#3 - #4	LB	0.032	0.63	2.71		3.34
1030	#5 - #6	"	0.027	0.63	2.26		2.89
1035	Vertical						
1040	#3 - #4	LB	0.040	0.63	3.39		4.02
1050	#5 - #6	"	0.032	0.63	2.71		3.34
1100	Horizontal joint reinforcing						
1105	Truss type						
1110	4" wide, 6" wall	LF	0.003	0.20	0.27		0.47
1120	6" wide, 8" wall	"	0.003	0.20	0.28		0.48
1130	8" wide, 10" wall	"	0.003	0.25	0.29		0.54
1140	10" wide, 12" wall	"	0.004	0.25	0.30		0.55
1150	12" wide, 14" wall	"	0.004	0.31	0.32		0.63
1155	Ladder type						
1160	4" wide, 6" wall	LF	0.003	0.15	0.27		0.42
1170	6" wide, 8" wall	"	0.003	0.17	0.28		0.45
1180	8" wide, 10" wall	"	0.003	0.18	0.29		0.47
1190	10" wide, 12" wall	"	0.003	0.21	0.29		0.50
2000	Rectangular wall ties						
2005	3/16" dia., galvanized						
2010	2" x 6"	EA	0.013	0.38	1.13		1.51
2020	2" x 8"	"	0.013	0.41	1.13		1.54
2040	2" x 10"	"	0.013	0.47	1.13		1.60
2050	2" x 12"	"	0.013	0.53	1.13		1.66
2060	4" x 6"	"	0.016	0.43	1.35		1.78
2070	4" x 8"	"	0.016	0.49	1.35		1.84
2080	4" x 10"	"	0.016	0.64	1.35		1.99
2090	4" x 12"	"	0.016	0.73	1.35		2.08
2095	1/4" dia., galvanized						
2100	2" x 6"	EA	0.013	0.71	1.13		1.84
2110	2" x 8"	"	0.013	0.80	1.13		1.93
2120	2" x 10"	"	0.013	0.90	1.13		2.03

MORTAR, GROUT AND ACCESSORIES

ID Code	Description / Component Descriptions	Output Unit of Meas.	Manhr / Unit	Material Cost	Labor Cost	Equipment Cost	Total Cost
04 - 05231	**MASONRY ACCESSORIES, Cont'd...**						**04 - 05231**
2130	2" x 12"	EA	0.013	1.04	1.13		2.17
2140	4" x 6"	"	0.016	0.82	1.35		2.17
2150	4" x 8"	"	0.016	0.90	1.35		2.25
2160	4" x 10"	"	0.016	1.04	1.35		2.39
2170	4" x 12"	"	0.016	1.09	1.35		2.44
2200	Z-type wall ties, galvanized						
2215	6" long						
2220	1/8" dia.	EA	0.013	0.33	1.13		1.46
2230	3/16" dia.	"	0.013	0.36	1.13		1.49
2240	1/4" dia.	"	0.013	0.38	1.13		1.51
2245	8" long						
2250	1/8" dia.	EA	0.013	0.36	1.13		1.49
2260	3/16" dia.	"	0.013	0.38	1.13		1.51
2270	1/4" dia.	"	0.013	0.41	1.13		1.54
2275	10" long						
2280	1/8" dia.	EA	0.013	0.38	1.13		1.51
2290	3/16" dia.	"	0.013	0.43	1.13		1.56
2300	1/4" dia.	"	0.013	0.49	1.13		1.62
3000	Dovetail anchor slots						
3015	Galvanized steel, filled						
3020	24 ga.	LF	0.020	0.82	1.69		2.51
3040	20 ga.	"	0.020	1.72	1.69		3.41
3060	16 oz. copper, foam filled	"	0.020	2.03	1.69		3.72
3100	Dovetail anchors						
3115	16 ga.						
3120	3-1/2" long	EA	0.013	0.25	1.13		1.38
3140	5-1/2" long	"	0.013	0.30	1.13		1.43
3150	12 ga.						
3160	3-1/2" long	EA	0.013	0.33	1.13		1.46
3180	5-1/2" long	"	0.013	0.55	1.13		1.68
3200	Dovetail, triangular galvanized ties, 12 ga.						
3220	3" x 3"	EA	0.013	0.56	1.13		1.69
3240	5" x 5"	"	0.013	0.60	1.13		1.73
3260	7" x 7"	"	0.013	0.68	1.13		1.81
3280	7" x 9"	"	0.013	0.72	1.13		1.85
3400	Brick anchors						
3420	Corrugated, 3-1/2" long						
3440	16 ga.	EA	0.013	0.23	1.13		1.36

MORTAR, GROUT AND ACCESSORIES

ID Code	Description — Component Descriptions	Output — Unit of Meas.	Output — Manhr / Unit	Unit Costs — Material Cost	Unit Costs — Labor Cost	Unit Costs — Equipment Cost	Unit Costs — Total Cost
04 - 05231	**MASONRY ACCESSORIES, Cont'd...**						**04 - 05231**
3460	12 ga.	EA	0.013	0.40	1.13		1.53
3500	Non-corrugated, 3-1/2" long						
3520	16 ga.	EA	0.013	0.32	1.13		1.45
3540	12 ga.	"	0.013	0.57	1.13		1.70
3580	Cavity wall anchors, corrugated, galvanized						
3600	5" long						
3620	16 ga.	EA	0.013	0.71	1.13		1.84
3640	12 ga.	"	0.013	1.06	1.13		2.19
3660	7" long						
3680	28 ga.	EA	0.013	0.78	1.13		1.91
3700	24 ga.	"	0.013	0.99	1.13		2.12
3720	22 ga.	"	0.013	1.01	1.13		2.14
3740	16 ga.	"	0.013	1.14	1.13		2.27
3800	Mesh ties, 16 ga., 3" wide						
3820	8" long	EA	0.013	0.95	1.13		2.08
3840	12" long	"	0.013	1.06	1.13		2.19
3860	20" long	"	0.013	1.46	1.13		2.59
3900	24" long	"	0.013	1.61	1.13		2.74
04 - 05232	**MASONRY CONTROL JOINTS**						**04 - 05232**
1000	Control joint, cross shaped PVC	LF	0.020	2.17	1.69		3.86
1010	Closed cell joint filler						
1020	1/2"	LF	0.020	0.38	1.69		2.07
1040	3/4"	"	0.020	0.77	1.69		2.46
1070	Rubber, for						
1080	4" wall	LF	0.020	2.50	1.69		4.19
1110	PVC, for						
1120	4" wall	LF	0.020	1.30	1.69		2.99
04 - 05235	**MASONRY FLASHING**						**04 - 05235**
0080	Through-wall flashing						
1000	5 oz. coated copper	SF	0.067	3.35	5.65		9.00
1020	0.030" elastomeric	"	0.053	1.10	4.52		5.62

UNIT MASONRY

ID Code	Description / Component Descriptions	Output		Unit Costs			
		Unit of Meas.	Manhr / Unit	Material Cost	Labor Cost	Equipment Cost	Total Cost
04 - 21131	**BRICK MASONRY**						**04 - 21131**
0100	Standard size brick, running bond						
1000	Face brick, red (6.4/sf)						
1020	Veneer	SF	0.133	5.39	11.25		16.75
1030	Cavity wall	"	0.114	5.39	9.68		15.00
1040	9" solid wall	"	0.229	10.75	19.25		30.00
1200	Common brick (6.4/sf)						
1210	Select common for veneers	SF	0.133	3.52	11.25		14.75
1215	Back-up						
1220	4" thick	SF	0.100	3.16	8.47		11.75
1230	8" thick	"	0.160	6.33	13.50		19.75
1300	Glazed brick (7.4/sf)						
1310	Veneer	SF	0.145	14.75	12.25		27.00
1400	Buff or gray face brick (6.4/sf)						
1410	Veneer	SF	0.133	6.27	11.25		17.50
1420	Cavity wall	"	0.114	6.27	9.68		16.00
1500	Jumbo or oversize brick (3/sf)						
1510	4" veneer	SF	0.080	4.12	6.78		11.00
1530	4" back-up	"	0.067	4.12	5.65		9.77
1540	8" back-up	"	0.114	4.78	9.68		14.50
1600	Norman brick, red face (4.5/sf)						
1620	4" veneer	SF	0.100	5.94	8.47		14.50
1640	Cavity wall	"	0.089	5.94	7.53		13.50
3000	Chimney, standard brick, including flue						
3020	16" x 16"	LF	0.800	25.25	68.00		93.00
3040	16" x 20"	"	0.800	43.00	68.00		110
3060	16" x 24"	"	0.800	45.75	68.00		110
3080	20" x 20"	"	1.000	35.75	85.00		120
3100	20" x 24"	"	1.000	48.25	85.00		130
3120	20" x 32"	"	1.143	54.00	97.00		150
4000	Window sill, face brick on edge	"	0.200	2.97	17.00		20.00

CONCRETE UNIT MASONRY

ID Code	Component Descriptions	Unit of Meas.	Manhr / Unit	Material Cost	Labor Cost	Equipment Cost	Total Cost
	Description	**Output**		**Unit Costs**			

04 - 22001   CONCRETE MASONRY UNITS   04 - 22001

ID Code	Component Descriptions	Unit of Meas.	Manhr / Unit	Material Cost	Labor Cost	Equipment Cost	Total Cost
0110	Hollow, load bearing						
0120	4"	SF	0.059	1.55	5.02		6.57
0140	6"	"	0.062	2.27	5.21		7.48
0160	8"	"	0.067	2.60	5.65		8.25
0180	10"	"	0.073	3.60	6.16		9.76
0190	12"	"	0.080	4.13	6.78		11.00
0280	Solid, load bearing						
0300	4"	SF	0.059	2.43	5.02		7.45
0320	6"	"	0.062	2.73	5.21		7.94
0340	8"	"	0.067	3.72	5.65		9.37
0360	10"	"	0.073	3.97	6.16		10.25
0380	12"	"	0.080	5.89	6.78		12.75
0480	Back-up block, 8" x 16"						
0490	2"	SF	0.046	1.63	3.87		5.50
0540	4"	"	0.047	1.69	3.98		5.67
0560	6"	"	0.050	2.48	4.23		6.71
0580	8"	"	0.053	2.85	4.52		7.37
0600	10"	"	0.057	3.94	4.84		8.78
0620	12"	"	0.062	4.52	5.21		9.73
0980	Foundation wall, 8" x 16"						
1000	6"	SF	0.057	2.48	4.84		7.32
1030	8"	"	0.062	2.85	5.21		8.06
1040	10"	"	0.067	3.93	5.65		9.58
1050	12"	"	0.073	4.52	6.16		10.75
1055	Solid						
1060	6"	SF	0.062	2.99	5.21		8.20
1070	8"	"	0.067	4.08	5.65		9.73
1080	10"	"	0.073	4.36	6.16		10.50
1100	12"	"	0.080	6.47	6.78		13.25
1480	Exterior, styrofoam inserts, std weight, 8" x 16"						
1500	6"	SF	0.062	4.35	5.21		9.56
1530	8"	"	0.067	4.70	5.65		10.25
1540	10"	"	0.073	6.09	6.16		12.25
1550	12"	"	0.080	8.36	6.78		15.25
1580	Lightweight						
1600	6"	SF	0.062	4.85	5.21		10.00
1660	8"	"	0.067	5.45	5.65		11.00
1680	10"	"	0.073	5.80	6.16		12.00

CONCRETE UNIT MASONRY

ID Code	Component Descriptions	Unit of Meas.	Manhr / Unit	Material Cost	Labor Cost	Equipment Cost	Total Cost
		Description				**Unit Costs**	
			Output				

ID Code	Component Descriptions	Unit of Meas.	Manhr / Unit	Material Cost	Labor Cost	Equipment Cost	Total Cost
04 - 22001	**CONCRETE MASONRY UNITS, Cont'd...**						**04 - 22001**
1700	12"	SF	0.080	7.65	6.78		14.50
1980	Acoustical slotted block						
2000	4"	SF	0.073	5.05	6.16		11.25
2020	6"	"	0.073	5.30	6.16		11.50
2040	8"	"	0.080	6.61	6.78		13.50
2050	Filled cavities						
2060	4"	SF	0.089	5.41	7.53		13.00
2070	6"	"	0.094	6.24	7.97		14.25
2080	8"	"	0.100	8.00	8.47		16.50
4000	Hollow, split face						
4020	4"	SF	0.059	3.46	5.02		8.48
4030	6"	"	0.062	4.01	5.21		9.22
4040	8"	"	0.067	4.21	5.65		9.86
4080	10"	"	0.073	4.71	6.16		10.75
4100	12"	"	0.080	5.03	6.78		11.75
4480	Split rib profile						
4500	4"	SF	0.073	4.21	6.16		10.25
4520	6"	"	0.073	4.89	6.16		11.00
4540	8"	"	0.080	5.32	6.78		12.00
4560	10"	"	0.080	5.82	6.78		12.50
4580	12"	"	0.080	6.31	6.78		13.00
5500	Solar screen concrete block						
5505	4" thick						
5510	6" x 6"	SF	0.178	3.89	15.00		19.00
5520	8" x 8"	"	0.160	4.65	13.50		18.25
5530	12" x 12"	"	0.123	4.76	10.50		15.25
5540	8" thick						
5550	8" x 16"	SF	0.114	4.76	9.68		14.50
9900	Vertical reinforcing						
9920	4' o.c., add 5% to labor						
9940	2'8" o.c., add 15% to labor						
9960	Interior partitions, add 10% to labor						

CONCRETE UNIT MASONRY

ID Code	Description / Component Descriptions	Output / Unit of Meas.	Manhr / Unit	Unit Costs / Material Cost	Labor Cost	Equipment Cost	Total Cost
04 - 22009	**BOND BEAMS & LINTELS**						**04 - 22009**
0980	Bond beam, no grout or reinforcement						
0990	8" x 16" x						
1000	4" thick	LF	0.062	1.46	5.21		6.67
1040	6" thick	"	0.064	2.24	5.42		7.66
1060	8" thick	"	0.067	2.56	5.65		8.21
1080	10" thick	"	0.070	3.17	5.89		9.06
1100	12" thick	"	0.073	3.60	6.16		9.76
6000	Beam lintel, no grout or reinforcement						
6010	8" x 16" x						
6020	10" thick	LF	0.080	5.86	6.78		12.75
6040	12" thick	"	0.089	7.44	7.53		15.00
6080	Precast masonry lintel						
7000	6 lf, 8" high x						
7020	4" thick	LF	0.133	5.36	11.25		16.50
7040	6" thick	"	0.133	6.85	11.25		18.00
7060	8" thick	"	0.145	7.75	12.25		20.00
7080	10" thick	"	0.145	9.25	12.25		21.50
7090	10 lf, 8" high x						
7100	4" thick	LF	0.080	6.74	6.78		13.50
7120	6" thick	"	0.080	8.31	6.78		15.00
7140	8" thick	"	0.089	9.25	7.53		16.75
7160	10" thick	"	0.089	12.50	7.53		20.00
8000	Steel angles and plates						
8010	Minimum	LB	0.011	1.09	0.96		2.05
8020	Maximum	"	0.020	1.61	1.69		3.30
8200	Various size angle lintels						
8205	1/4" stock						
8210	3" x 3"	LF	0.050	5.65	4.23		9.88
8220	3" x 3-1/2"	"	0.050	6.23	4.23		10.50
8225	3/8" stock						
8230	3" x 4"	LF	0.050	9.82	4.23		14.00
8240	3-1/2" x 4"	"	0.050	10.25	4.23		14.50
8250	4" x 4"	"	0.050	11.25	4.23		15.50
8260	5" x 3-1/2"	"	0.050	12.00	4.23		16.25
8262	6" x 3-1/2"	"	0.050	13.50	4.23		17.75
8265	1/2" stock						
8280	6" x 4"	LF	0.050	15.00	4.23		19.25

GLASS UNIT MASONRY

ID Code	Component Descriptions	Unit of Meas.	Manhr / Unit	Material Cost	Labor Cost	Equipment Cost	Total Cost
	Description	**Output**		**Unit Costs**			
04 - 23001	**GLASS BLOCK**						**04 - 23001**
1000	Glass block, 4" thick						
1040	6" x 6"	SF	0.267	30.25	22.50		53.00
1060	8" x 8"	"	0.200	19.00	17.00		36.00
1080	12" x 12"	"	0.160	24.25	13.50		37.75

STONE

ID Code	Component Descriptions	Unit of Meas.	Manhr / Unit	Material Cost	Labor Cost	Equipment Cost	Total Cost
04 - 43001	**STONE**						**04 - 43001**
0160	Rubble stone						
0180	Walls set in mortar						
0200	8" thick	SF	0.200	15.00	17.00		32.00
0220	12" thick	"	0.320	18.25	27.00		45.25
0420	18" thick	"	0.400	24.00	34.00		58.00
0440	24" thick	"	0.533	30.25	45.25		76.00
0445	Dry set wall						
0450	8" thick	SF	0.133	17.00	11.25		28.25
0455	12" thick	"	0.200	19.00	17.00		36.00
0460	18" thick	"	0.267	26.50	22.50		49.00
0465	24" thick	"	0.320	32.25	27.00		59.00
1525	Thresholds, 7/8" thick, 3' long, 4" to 6" wide						
1530	Plain	EA	0.667	27.00	57.00		84.00
1540	Beveled	"	0.667	30.00	57.00		87.00
1545	Window sill						
1550	6" wide, 2" thick	LF	0.320	15.25	27.00		42.25
1555	Stools						
1560	5" wide, 7/8" thick	LF	0.320	22.25	27.00		49.25
1800	Granite veneer facing panels, polished						
1810	7/8" thick						
1820	Black	SF	0.320	51.00	27.00		78.00
1840	Gray	"	0.320	40.00	27.00		67.00
2000	Slate, panels						
2010	1" thick	SF	0.320	25.00	27.00		52.00
2030	Sills or stools						
2040	1" thick						
2060	6" wide	LF	0.320	11.75	27.00		38.75
2080	10" wide	"	0.348	19.00	29.50		48.50

REFRACTORIES

ID Code	Description		Output		Unit Costs			
	Component Descriptions		Unit of Meas.	Manhr / Unit	Material Cost	Labor Cost	Equipment Cost	Total Cost
04 - 51001			**FLUE LINERS**					**04 - 51001**
1000	Flue liners							
1020	Rectangular							
1040	8" x 12"		LF	0.133	7.91	11.25		19.25
1060	12" x 12"		"	0.145	9.87	12.25		22.00
1080	12" x 18"		"	0.160	17.50	13.50		31.00
1100	16" x 16"		"	0.178	18.75	15.00		33.75
1120	18" x 18"		"	0.190	23.25	16.25		39.50
1140	20" x 20"		"	0.200	38.75	17.00		56.00
1170	24" x 24"		"	0.229	46.50	19.25		66.00
1200	Round							
1220	18" dia.		LF	0.190	35.75	16.25		52.00
1240	24" dia.		"	0.229	71.00	19.25		90.00

DIVISION 05
METALS

METAL FASTENINGS

ID Code	Component Descriptions	Unit of Meas.	Manhr / Unit	Material Cost	Labor Cost	Equipment Cost	Total Cost
	Description	**Output**		**Unit Costs**			
05 - 05231	**STRUCTURAL WELDING**						**05 - 05231**
0080	Welding						
0100	Single pass						
0120	1/8"	LF	0.040	0.43	3.59		4.02
0140	3/16"	"	0.053	0.72	4.79		5.51
0160	1/4"	"	0.067	1.01	5.99		7.00
05 - 05239	**METAL FASTENINGS**						**05 - 05239**
1000	Anchor bolts, material only						
1020	3/8" x						
1040	8" long	EA					1.67
1080	12" long	"					2.42
1090	1/2" x						
1100	8" long	EA					2.64
1140	12" long	"					3.41
1170	5/8" x						
1180	8" long	EA					5.22
1220	12" long	"					6.43
1270	3/4" x						
1280	8" long	EA					6.38
1300	12" long	"					9.13
4480	Non-drilling anchor						
4500	1/4"	EA					0.85
4540	3/8"	"					1.06
4560	1/2"	"					1.62
7000	Self-drilling anchor						
7020	1/4"	EA					2.14
7060	3/8"	"					3.22
7080	1/2"	"					4.29
05 - 05240	**METAL LINTELS**						**05 - 05240**
0080	Lintels, steel						
0100	Plain	LB	0.020	1.98	1.79		3.77
0120	Galvanized	"	0.020	2.39	1.79		4.18

STRUCTURAL STEEL

| ID Code | Description | Output | | Unit Costs | | | |
	Component Descriptions	Unit of Meas.	Manhr / Unit	Material Cost	Labor Cost	Equipment Cost	Total Cost
05 - 12001	**BEAMS, GIRDERS, COLUMNS, TRUSSES**						**05 - 12001**
0100	Beams and girders, A-36						
0120	Welded	TON	4.800	3,950	430	550	4,930
0140	Bolted	"	4.364	3,840	390	500	4,730
0180	Columns						
0185	Pipe						
0190	6" dia.	LB	0.005	2.02	0.42	0.55	2.99
1300	Structural tube						
1310	6" square						
1320	Light sections	TON	9.600	4,710	850	1,100	6,660

COLD FORMED FRAMING

ID Code	Description	Output		Unit Costs			
05 - 41001	**METAL FRAMING**						**05 - 41001**
0100	Furring channel, galvanized						
0110	Beams and columns, 3/4"						
0120	12" o.c.	SF	0.080	0.48	7.19		7.67
0140	16" o.c.	"	0.073	0.37	6.53		6.90
0150	Walls, 3/4"						
0160	12" o.c.	SF	0.040	0.48	3.59		4.07
0170	16" o.c.	"	0.033	0.37	2.99		3.36
0172	24" o.c.	"	0.027	0.26	2.39		2.65
0173	1-1/2"						
0174	12" o.c.	SF	0.040	0.79	3.59		4.38
0175	16" o.c.	"	0.033	0.60	2.99		3.59
0176	24" o.c.	"	0.027	0.41	2.39		2.80
2000	Stud, load bearing						
2020	16" o.c.						
2040	12 ga.						
2060	2-1/2"	SF	0.036	1.78	3.19		4.97
2080	3-5/8"	"	0.036	2.10	3.19		5.29
2100	4"	"	0.036	2.21	3.19		5.40
2120	6"	"	0.040	2.75	3.59		6.34
2140	8"	"	0.040	3.35	3.59		6.94
2160	12"	"	0.044	4.42	3.99		8.41
2200	14 ga.						
2220	2-1/2"	SF	0.036	1.46	3.19		4.65
2240	3-5/8"	"	0.036	1.72	3.19		4.91
2260	4"	"	0.036	1.81	3.19		5.00

COLD FORMED FRAMING

ID Code	Component Descriptions	Unit of Meas.	Manhr / Unit	Material Cost	Labor Cost	Equipment Cost	Total Cost
	Description	**Output**		**Unit Costs**			

ID Code	Component Descriptions	Unit of Meas.	Manhr / Unit	Material Cost	Labor Cost	Equipment Cost	Total Cost
05 - 41001	**METAL FRAMING, Cont'd...**						**05 - 41001**
2280	6"	SF	0.040	2.25	3.59		5.84
2300	8"	"	0.040	2.75	3.59		6.34
2320	12"	"	0.044	3.63	3.99		7.62
2340	16 ga.						
2360	2-1/2"	SF	0.036	1.33	3.19		4.52
2380	3-5/8"	"	0.036	1.57	3.19		4.76
2400	4"	"	0.036	1.63	3.19		4.82
2420	6"	"	0.040	2.05	3.59		5.64
2440	8"	"	0.040	2.64	3.59		6.23
2460	12"	"	0.044	3.30	3.99		7.29
2480	18 ga.						
2500	2-1/2"	SF	0.036	1.08	3.19		4.27
2520	3-5/8"	"	0.036	1.33	3.19		4.52
2540	4"	"	0.036	1.39	3.19		4.58
2560	6"	"	0.040	1.76	3.59		5.35
2580	8"	"	0.040	2.12	3.59		5.71
2600	12"	"	0.044	2.64	3.99		6.63
2620	20 ga.						
2640	2-1/2"	SF	0.036	0.72	3.19		3.91
2660	3-5/8"	"	0.036	0.87	3.19		4.06
2680	4"	"	0.036	0.95	3.19		4.14
2700	6"	"	0.040	1.16	3.59		4.75
2720	8"	"	0.040	1.38	3.59		4.97
2740	12"	"	0.044	1.65	3.99		5.64
2760	24" o.c.						
2780	12 ga.						
2800	2-1/2"	SF	0.031	1.24	2.76		4.00
2820	3-5/8"	"	0.031	1.47	2.76		4.23
2840	4"	"	0.031	1.55	2.76		4.31
2860	6"	"	0.033	1.92	2.99		4.91
2880	8"	"	0.033	2.34	2.99		5.33
2900	12"	"	0.036	3.10	3.26		6.36
2920	14 ga.						
2940	2-1/2"	SF	0.031	1.02	2.76		3.78
2960	3-5/8"	"	0.031	1.20	2.76		3.96
2980	4"	"	0.031	1.27	2.76		4.03
3000	6"	"	0.033	1.57	2.99		4.56
3020	8"	"	0.033	1.92	2.99		4.91

COLD FORMED FRAMING

ID Code	Component Descriptions	Unit of Meas.	Manhr / Unit	Material Cost	Labor Cost	Equipment Cost	Total Cost
05 - 41001	**METAL FRAMING, Cont'd...**						**05 - 41001**
3040	12"	SF	0.036	2.54	3.26		5.80
3060	16 ga.						
3080	2-1/2"	SF	0.031	0.91	2.76		3.67
3100	3-5/8"	"	0.031	1.08	2.76		3.84
3120	4"	"	0.031	1.15	2.76		3.91
3140	6"	"	0.033	1.39	2.99		4.38
3160	8"	"	0.033	1.76	2.99		4.75
3180	12"	"	0.036	1.76	3.26		5.02
3200	18 ga.						
3220	2-1/2"	SF	0.031	0.72	2.76		3.48
3240	3-5/8"	"	0.031	0.84	2.76		3.60
3260	4"	"	0.031	0.91	2.76		3.67
3280	6"	"	0.033	1.15	2.99		4.14
3300	8"	"	0.033	1.39	2.99		4.38
3320	12"	"	0.036	1.39	3.26		4.65
3340	20 ga.						
3360	2-1/2"	SF	0.031	0.46	2.76		3.22
3380	3-5/8"	"	0.031	0.51	2.76		3.27
3400	4"	"	0.031	0.57	2.76		3.33
3420	6"	"	0.033	0.75	2.99		3.74
3440	8"	"	0.033	0.92	2.99		3.91
3460	12"	"	0.036	0.92	3.26		4.18
8020	Track, load bearing, 12 ga.						
8040	2-1/2"	LF	0.012	1.78	1.10		2.88
8060	3-5/8"	"	0.012	2.10	1.10		3.20
8080	4"	"	0.012	2.21	1.10		3.31
8100	6"	"	0.013	2.75	1.14		3.89
8120	8"	"	0.013	3.35	1.14		4.49
8140	12"	"	0.013	4.42	1.17		5.59
8160	14 ga.						
8180	2-1/2"	LF	0.012	1.46	1.10		2.56
8200	3-5/8"	"	0.012	1.72	1.10		2.82
8220	4"	"	0.012	1.81	1.10		2.91
8240	6"	"	0.013	2.25	1.14		3.39
8260	8"	"	0.013	2.75	1.14		3.89
8280	12"	"	0.013	3.63	1.17		4.80
8300	16 ga.						
8320	2-1/2"	LF	0.012	1.33	1.10		2.43

COLD FORMED FRAMING

ID Code	Component Descriptions	Unit of Meas.	Manhr / Unit	Material Cost	Labor Cost	Equipment Cost	Total Cost
	Description	**Output**		**Unit Costs**			
05 - 41001	**METAL FRAMING, Cont'd...**						**05 - 41001**
8340	3-5/8"	LF	0.012	1.57	1.10		2.67
8360	4"	"	0.012	1.63	1.10		2.73
8380	6"	"	0.013	2.05	1.14		3.19
8400	8"	"	0.013	2.64	1.14		3.78
8420	12"	"	0.013	3.30	1.17		4.47
8440	18 ga.						
8460	2-1/2"	LF	0.012	1.08	1.10		2.18
8480	3-5/8"	"	0.012	1.33	1.10		2.43
8500	4"	"	0.012	1.39	1.10		2.49
8520	6"	"	0.013	1.76	1.14		2.90
8540	8"	"	0.013	2.12	1.14		3.26
8560	12"	"	0.013	2.64	1.17		3.81
8600	20 ga.						
8620	2-1/2"	LF	0.012	0.72	1.10		1.82
8640	3-5/8"	"	0.012	0.87	1.10		1.97
8660	4"	"	0.012	0.95	1.10		2.05
8680	6"	"	0.013	1.16	1.14		2.30
8700	8"	"	0.013	1.38	1.14		2.52
8720	12"	"	0.013	1.65	1.17		2.82

METAL FABRICATIONS

ID Code	Component Descriptions	Unit of Meas.	Manhr / Unit	Material Cost	Labor Cost	Equipment Cost	Total Cost
05 - 52131	**RAILINGS**						**05 - 52131**
0080	Railing, pipe						
0090	1-1/4" diameter, welded steel						
0095	2-rail						
0100	Primed	LF	0.160	41.25	14.50		56.00
0120	Galvanized	"	0.160	53.00	14.50		68.00
0130	3-rail						
0140	Primed	LF	0.200	53.00	18.00		71.00
0160	Galvanized	"	0.200	69.00	18.00		87.00
0170	Wall mounted, single rail, welded steel						
0180	Primed	LF	0.123	27.50	11.00		38.50
0200	Galvanized	"	0.123	35.75	11.00		46.75
0270	Wall mounted, single rail, welded steel						
0280	Primed	LF	0.123	28.50	11.00		39.50
0300	Galvanized	"	0.123	37.00	11.00		48.00
1075	Wall mounted, single rail, welded steel						

METAL FABRICATIONS

ID Code	Description		Output		Unit Costs			
	Component Descriptions		Unit of Meas.	Manhr / Unit	Material Cost	Labor Cost	Equipment Cost	Total Cost
05 - 52131		**RAILINGS, Cont'd...**						**05 - 52131**
1080	Primed		LF	0.133	30.75	12.00		42.75
1100	Galvanized		"	0.133	40.00	12.00		52.00

MISC. FABRICATIONS

05 - 73001	**ORNAMENTAL METAL**						**05 - 73001**
1030	Railings, square bars, 6" o.c., shaped top rails						
1040	Steel	LF	0.400	99.00	36.00		140
1060	Aluminum	"	0.400	120	36.00		160
1080	Bronze	"	0.533	250	48.00		300
1100	Stainless steel	"	0.533	250	48.00		300
1200	Laminated metal or wood handrails						
1220	2-1/2" round or oval shape	LF	0.400	290	36.00		330

DIVISION 06
WOOD AND PLASTICS

FASTENERS AND ADHESIVES

ID Code	Description / Component Descriptions	Output		Unit Costs			
		Unit of Meas.	Manhr / Unit	Material Cost	Labor Cost	Equipment Cost	Total Cost
06 - 05231	**ACCESSORIES**						**06 - 05231**
0080	Column/post base, cast aluminum						
0100	4" x 4"	EA	0.200	20.75	17.75		38.50
0120	6" x 6"	"	0.200	29.00	17.75		46.75
0130	Bridging, metal, per pair						
0140	12" o.c.	EA	0.080	2.92	7.10		10.00
0160	16" o.c.	"	0.073	2.69	6.45		9.14
1000	Anchors						
1020	Bolts, threaded two ends, with nuts and washers						
1030	1/2" dia.						
1040	4" long	EA	0.050	3.75	4.44		8.19
1060	7-1/2" long	"	0.050	4.37	4.44		8.81
1070	3/4" dia.						
1080	7-1/2" long	EA	0.050	7.39	4.44		11.75
1100	15" long	"	0.050	11.25	4.44		15.75
1200	Framing anchors						
1202	10 gauge	EA	0.067	1.60	5.92		7.52
1210	Bolts, carriage						
1212	1/4 x 4	EA	0.080	0.85	7.10		7.95
1214	5/16 x 6	"	0.084	1.93	7.47		9.40
1216	3/8 x 6	"	0.084	3.91	7.47		11.50
1218	1/2 x 6	"	0.084	5.45	7.47		13.00
1240	Joist and beam hangers						
1250	18 ga.						
1260	2 x 4	EA	0.080	1.61	7.10		8.71
1280	2 x 6	"	0.080	1.93	7.10		9.03
1282	2 x 8	"	0.080	2.25	7.10		9.35
1284	2 x 10	"	0.089	2.42	7.89		10.25
1286	2 x 12	"	0.100	4.07	8.88		13.00
1288	16 ga.						
1290	3 x 6	EA	0.089	6.08	7.89		14.00
1292	3 x 8	"	0.089	7.38	7.89		15.25
1300	3 x 10	"	0.094	8.33	8.35		16.75
1302	3 x 12	"	0.107	9.38	9.47		18.75
1304	3 x 14	"	0.114	10.25	10.25		20.50
1320	4 x 6	"	0.089	9.47	7.89		17.25
1322	4 x 8	"	0.089	11.00	7.89		19.00
1324	4 x 10	"	0.094	12.75	8.35		21.00
1326	4 x 12	"	0.107	16.25	9.47		25.75

FASTENERS AND ADHESIVES

ID Code	Description Component Descriptions	Output Unit of Meas.	Manhr / Unit	Unit Costs Material Cost	Labor Cost	Equipment Cost	Total Cost
06 - 05231	**ACCESSORIES, Cont'd...**						**06 - 05231**
1328	4 x 14	EA	0.114	17.25	10.25		27.50
1520	Rafter anchors, 18 ga., 1-1/2" wide						
1540	5-1/4" long	EA	0.067	1.19	5.92		7.11
1560	10-3/4" long	"	0.067	1.76	5.92		7.68
1600	Shear plates						
1620	2-5/8" dia.	EA	0.062	4.23	5.46		9.69
1640	4" dia.	"	0.067	8.78	5.92		14.75
1700	Sill anchors						
1720	Embedded in concrete	EA	0.080	3.10	7.10		10.25
1800	Split rings						
1820	2-1/2" dia.	EA	0.089	2.55	7.89		10.50
1840	4" dia.	"	0.100	4.70	8.88		13.50
1900	Strap ties, 14 ga., 1-3/8" wide						
1920	12" long	EA	0.067	3.19	5.92		9.11
1940	18" long	"	0.073	3.43	6.45		9.88
1960	24" long	"	0.080	5.11	7.10		12.25
1980	36" long	"	0.089	7.02	7.89		15.00
2000	Toothed rings						
2020	2-5/8" dia.	EA	0.133	2.95	11.75		14.75
2040	4" dia.	"	0.160	3.43	14.25		17.75
06 - 05731	**WOOD TREATMENT**						**06 - 05731**
1000	Creosote preservative treatment						
1020	8 lb/cf	BF					0.74
1040	10 lb/cf	"					0.88
1060	Salt preservative treatment						
1070	Oil borne						
1080	Minimum	BF					0.67
1100	Maximum	"					0.95
1120	Water borne						
1140	Minimum	BF					0.47
1150	Maximum	"					0.74
1200	Fire retardant treatment						
1220	Minimum	BF					0.95
1240	Maximum	"					1.15
1300	Kiln dried, softwood, add to framing costs						
1320	1" thick	BF					0.34
1340	2" thick	"					0.47

FASTENERS AND ADHESIVES

ID Code	Description	Output		Unit Costs			
	Component Descriptions	Unit of Meas.	Manhr / Unit	Material Cost	Labor Cost	Equipment Cost	Total Cost
06 - 05731	**WOOD TREATMENT, Cont'd...**						**06 - 05731**
1360	3" thick	BF					0.60
1380	4" thick	"					0.74

ROUGH CARPENTRY

06 - 11001	**BLOCKING**						**06 - 11001**
1100	Steel construction						
1105	Walls						
1110	2x4	LF	0.053	1.06	4.73		5.79
1120	2x6	"	0.062	1.59	5.46		7.05
1130	2x8	"	0.067	2.12	5.92		8.04
1140	2x10	"	0.073	2.66	6.45		9.11
1150	2x12	"	0.080	3.25	7.10		10.25
1160	Ceilings						
1170	2x4	LF	0.062	1.06	5.46		6.52
1180	2x6	"	0.073	1.59	6.45		8.04
1190	2x8	"	0.080	2.12	7.10		9.22
1200	2x10	"	0.089	2.66	7.89		10.50
1210	2x12	"	0.100	3.25	8.88		12.25
1215	Wood construction						
1220	Walls						
1230	2x4	LF	0.044	1.06	3.94		5.00
1240	2x6	"	0.050	1.59	4.44		6.03
1250	2x8	"	0.053	2.12	4.73		6.85
1260	2x10	"	0.057	2.66	5.07		7.73
1270	2x12	"	0.062	3.25	5.46		8.71
1280	Ceilings						
1290	2x4	LF	0.050	1.06	4.44		5.50
1300	2x6	"	0.057	1.59	5.07		6.66
1310	2x8	"	0.062	2.12	5.46		7.58
1320	2x10	"	0.067	2.66	5.92		8.58
1330	2x12	"	0.073	3.25	6.45		9.70

ROUGH CARPENTRY

ID Code	Component Descriptions	Unit of Meas.	Manhr / Unit	Material Cost	Labor Cost	Equipment Cost	Total Cost
06 - 11002	**CEILING FRAMING**						**06 - 11002**
1000	Ceiling joists						
1010	12" o.c.						
1020	2x4	SF	0.019	1.06	1.69		2.75
1030	2x6	"	0.020	1.59	1.77		3.36
1040	2x8	"	0.021	2.12	1.86		3.98
1050	2x10	"	0.022	2.66	1.97		4.63
1060	2x12	"	0.024	3.25	2.08		5.33
1070	16" o.c.						
1080	2x4	SF	0.015	1.04	1.36		2.40
1090	2x6	"	0.016	1.19	1.42		2.61
1100	2x8	"	0.017	1.59	1.48		3.07
1110	2x10	"	0.017	1.99	1.54		3.53
1120	2x12	"	0.018	2.44	1.61		4.05
1130	24" o.c.						
1140	2x4	SF	0.013	0.53	1.12		1.65
1150	2x6	"	0.013	0.82	1.18		2.00
1160	2x8	"	0.014	1.06	1.24		2.30
1170	2x10	"	0.015	1.33	1.31		2.64
1180	2x12	"	0.016	1.62	1.39		3.01
1200	Headers and nailers						
1210	2x4	LF	0.026	1.06	2.29		3.35
1220	2x6	"	0.027	1.59	2.36		3.95
1230	2x8	"	0.029	2.12	2.53		4.65
1240	2x10	"	0.031	2.66	2.73		5.39
1250	2x12	"	0.033	3.25	2.96		6.21
1300	Sister joists for ceilings						
1310	2x4	LF	0.057	1.06	5.07		6.13
1320	2x6	"	0.067	1.59	5.92		7.51
1330	2x8	"	0.080	2.12	7.10		9.22
1340	2x10	"	0.100	2.66	8.88		11.50
1350	2x12	"	0.133	3.25	11.75		15.00
06 - 11003	**FLOOR FRAMING**						**06 - 11003**
1000	Floor joists						
1010	12" o.c.						
1020	2x6	SF	0.016	1.59	1.42		3.01
1030	2x8	"	0.016	2.12	1.44		3.56
1040	2x10	"	0.017	2.66	1.48		4.14

ROUGH CARPENTRY

ID Code	Component Descriptions	Unit of Meas.	Manhr / Unit	Material Cost	Labor Cost	Equipment Cost	Total Cost
06 - 11003	**FLOOR FRAMING, Cont'd...**						**06 - 11003**
1050	2x12	SF	0.017	3.25	1.54		4.79
1060	2x14	"	0.017	3.79	1.48		5.27
1070	3x6	"	0.017	4.12	1.51		5.63
1080	3x8	"	0.017	5.50	1.54		7.04
1090	3x10	"	0.018	6.87	1.61		8.48
1100	3x12	"	0.019	8.25	1.69		9.94
1120	3x14	"	0.020	9.62	1.77		11.50
1130	4x6	"	0.017	3.63	1.48		5.11
1140	4x8	"	0.017	4.84	1.54		6.38
1150	4x10	"	0.018	6.05	1.61		7.66
1160	4x12	"	0.019	7.26	1.69		8.95
1170	4x14	"	0.020	8.47	1.77		10.25
1180	16" o.c.						
1190	2x6	SF	0.013	1.19	1.18		2.37
1200	2x8	"	0.014	1.59	1.20		2.79
1220	2x10	"	0.014	1.99	1.22		3.21
1230	2x12	"	0.014	2.44	1.26		3.70
1240	2x14	"	0.015	2.83	1.31		4.14
1250	3x6	"	0.014	1.98	1.22		3.20
1260	3x8	"	0.014	2.64	1.26		3.90
1270	3x10	"	0.015	3.30	1.31		4.61
1280	3x12	"	0.015	3.96	1.36		5.32
1290	3x14	"	0.016	4.62	1.42		6.04
1300	4x6	"	0.014	2.64	1.22		3.86
1310	4x8	"	0.014	3.52	1.26		4.78
1320	4x10	"	0.015	4.40	1.31		5.71
1330	4x12	"	0.015	5.28	1.36		6.64
1340	4x14	"	0.016	6.16	1.42		7.58
2000	Sister joists for floors						
2010	2x4	LF	0.050	1.06	4.44		5.50
2020	2x6	"	0.057	1.59	5.07		6.66
2030	2x8	"	0.067	2.12	5.92		8.04
2040	2x10	"	0.080	2.66	7.10		9.76
2050	2x12	"	0.100	3.25	8.88		12.25
2060	3x6	"	0.080	4.12	7.10		11.25
2070	3x8	"	0.089	5.50	7.89		13.50
2080	3x10	"	0.100	6.87	8.88		15.75
2090	3x12	"	0.114	8.25	10.25		18.50

ROUGH CARPENTRY

ID Code	Description	Unit of Meas.	Manhr / Unit	Material Cost	Labor Cost	Equipment Cost	Total Cost
	Component Descriptions						
06 - 11003	**FLOOR FRAMING, Cont'd...**						**06 - 11003**
2100	4x6	LF	0.080	3.63	7.10		10.75
2110	4x8	"	0.089	4.84	7.89		12.75
2120	4x10	"	0.100	6.05	8.88		15.00
2130	4x12	"	0.114	7.26	10.25		17.50
06 - 11004	**FURRING**						**06 - 11004**
1100	Furring, wood strips						
1102	Walls						
1105	On masonry or concrete walls						
1107	1x2 furring						
1110	12" o.c.	SF	0.025	0.46	2.22		2.68
1120	16" o.c.	"	0.023	0.35	2.02		2.37
1130	24" o.c.	"	0.021	0.23	1.86		2.09
1135	1x3 furring						
1140	12" o.c.	SF	0.025	0.69	2.22		2.91
1150	16" o.c.	"	0.023	0.51	2.02		2.53
1160	24" o.c.	"	0.021	0.35	1.86		2.21
1165	On wood walls						
1167	1x2 furring						
1170	12" o.c.	SF	0.018	0.46	1.57		2.03
1180	16" o.c.	"	0.016	0.35	1.42		1.77
1190	24" o.c.	"	0.015	0.23	1.29		1.52
1195	1x3 furring						
1200	12" o.c.	SF	0.018	0.69	1.57		2.26
1210	16" o.c.	"	0.016	0.51	1.42		1.93
1220	24" o.c.	"	0.015	0.35	1.29		1.64
1224	Ceilings						
1226	On masonry or concrete ceilings						
1228	1x2 furring						
1230	12" o.c.	SF	0.044	0.46	3.94		4.40
1240	16" o.c.	"	0.040	0.35	3.55		3.90
1250	24" o.c.	"	0.036	0.23	3.22		3.45
1254	1x3 furring						
1260	12" o.c.	SF	0.044	0.69	3.94		4.63
1270	16" o.c.	"	0.040	0.51	3.55		4.06
1280	24" o.c.	"	0.036	0.35	3.22		3.57
1286	On wood ceilings						
1288	1x2 furring						

ROUGH CARPENTRY

ID Code	Component Descriptions	Unit of Meas.	Manhr / Unit	Material Cost	Labor Cost	Equipment Cost	Total Cost
	Description	**Output**		**Unit Costs**			
06 - 11004	**FURRING, Cont'd...**						**06 - 11004**
1290	12" o.c.	SF	0.030	0.46	2.63		3.09
1300	16" o.c.	"	0.027	0.35	2.36		2.71
1310	24" o.c.	"	0.024	0.23	2.15		2.38
1316	1x3						
1320	12" o.c.	SF	0.030	0.69	2.63		3.32
1330	16" o.c.	"	0.027	0.51	2.36		2.87
1340	24" o.c.	"	0.024	0.35	2.15		2.50
06 - 11005	**ROOF FRAMING**						**06 - 11005**
1000	Roof framing						
1005	Rafters, gable end						
1008	0-2 pitch (flat to 2-in-12)						
1010	12" o.c.						
1020	2x4	SF	0.017	1.06	1.48		2.54
1030	2x6	"	0.017	1.59	1.54		3.13
1040	2x8	"	0.018	2.12	1.61		3.73
1050	2x10	"	0.019	2.66	1.69		4.35
1060	2x12	"	0.020	3.25	1.77		5.02
1070	16" o.c.						
1080	2x6	SF	0.014	1.19	1.26		2.45
1090	2x8	"	0.015	1.59	1.31		2.90
1100	2x10	"	0.015	2.00	1.36		3.36
1110	2x12	"	0.016	2.44	1.42		3.86
1120	24" o.c.						
1130	2x6	SF	0.012	0.80	1.07		1.87
1140	2x8	"	0.013	1.06	1.11		2.17
1150	2x10	"	0.013	1.33	1.14		2.47
1160	2x12	"	0.013	1.62	1.18		2.80
1165	4-6 pitch (4-in-12 to 6-in-12)						
1170	12" o.c.						
1175	2x4	SF	0.017	1.06	1.54		2.60
1180	2x6	"	0.018	1.59	1.61		3.20
1190	2x8	"	0.019	2.12	1.69		3.81
1200	2x10	"	0.020	2.66	1.77		4.43
1210	2x12	"	0.021	3.25	1.86		5.11
1220	16" o.c.						
1230	2x6	SF	0.015	1.19	1.31		2.50
1240	2x8	"	0.015	1.59	1.36		2.95

ROUGH CARPENTRY

ID Code	Description / Component Descriptions	Unit of Meas.	Manhr / Unit	Material Cost	Labor Cost	Equipment Cost	Total Cost
06 - 11005	**ROOF FRAMING, Cont'd...**						**06 - 11005**
1250	2x10	SF	0.016	2.00	1.42		3.42
1260	2x12	"	0.017	2.44	1.48		3.92
1270	24" o.c.						
1280	2x6	SF	0.013	0.80	1.11		1.91
1290	2x8	"	0.013	1.06	1.14		2.20
1300	2x10	"	0.014	1.33	1.22		2.55
1310	2x12	"	0.015	1.62	1.36		2.98
1315	8-12 pitch (8-in-12 to 12-in-12)						
1320	12" o.c.						
1330	2x4	SF	0.018	1.15	1.61		2.76
1340	2x6	"	0.019	1.72	1.69		3.41
1350	2x8	"	0.020	2.29	1.77		4.06
1360	2x10	"	0.021	2.87	1.86		4.73
1370	2x12	"	0.022	3.51	1.97		5.48
1380	16" o.c.						
1390	2x6	SF	0.015	1.29	1.36		2.65
1400	2x8	"	0.016	1.72	1.42		3.14
1410	2x10	"	0.017	2.16	1.48		3.64
1420	2x12	"	0.017	2.63	1.54		4.17
1430	24" o.c.						
1440	2x6	SF	0.013	0.86	1.14		2.00
1450	2x8	"	0.013	1.15	1.18		2.33
1460	2x10	"	0.014	1.43	1.22		2.65
1470	2x12	"	0.014	1.75	1.26		3.01
2000	Ridge boards						
2010	2x6	LF	0.040	1.59	3.55		5.14
2020	2x8	"	0.044	2.12	3.94		6.06
2030	2x10	"	0.050	2.66	4.44		7.10
2040	2x12	"	0.057	3.25	5.07		8.32
3000	Hip rafters						
3010	2x6	LF	0.029	1.59	2.53		4.12
3020	2x8	"	0.030	2.12	2.63		4.75
3030	2x10	"	0.031	2.66	2.73		5.39
3040	2x12	"	0.032	3.25	2.84		6.09
3180	Jack rafters						
3190	4-6 pitch (4-in-12 to 6-in-12)						
3200	16" o.c.						
3210	2x6	SF	0.024	1.19	2.08		3.27

ROUGH CARPENTRY

	Description	Output		Unit Costs			
ID Code	Component Descriptions	Unit of Meas.	Manhr / Unit	Material Cost	Labor Cost	Equipment Cost	Total Cost
06 - 11005		**ROOF FRAMING, Cont'd...**					**06 - 11005**
3220	2x8	SF	0.024	1.59	2.15		3.74
3230	2x10	"	0.026	2.00	2.29		4.29
3240	2x12	"	0.027	2.44	2.36		4.80
3250	24" o.c.						
3260	2x6	SF	0.018	0.80	1.61		2.41
3270	2x8	"	0.019	1.06	1.65		2.71
3280	2x10	"	0.020	1.33	1.73		3.06
3290	2x12	"	0.020	1.62	1.77		3.39
3295	8-12 pitch (8-in-12 to 12-in-12)						
3300	16" o.c.						
3310	2x6	SF	0.025	1.29	2.22		3.51
3320	2x8	"	0.026	1.72	2.29		4.01
3330	2x10	"	0.027	2.16	2.36		4.52
3340	2x12	"	0.028	2.63	2.44		5.07
3350	24" o.c.						
3360	2x6	SF	0.019	0.86	1.69		2.55
3370	2x8	"	0.020	1.15	1.73		2.88
3380	2x10	"	0.020	1.43	1.77		3.20
3390	2x12	"	0.021	1.75	1.82		3.57
4980	Sister rafters						
5000	2x4	LF	0.057	1.06	5.07		6.13
5010	2x6	"	0.067	1.59	5.92		7.51
5020	2x8	"	0.080	2.12	7.10		9.22
5030	2x10	"	0.100	2.66	8.88		11.50
5040	2x12	"	0.133	3.25	11.75		15.00
5050	Fascia boards						
5060	2x4	LF	0.040	1.16	3.55		4.71
5070	2x6	"	0.040	1.76	3.55		5.31
5080	2x8	"	0.044	2.34	3.94		6.28
5090	2x10	"	0.044	2.93	3.94		6.87
5100	2x12	"	0.050	3.52	4.44		7.96
7980	Cant strips						
7985	Fiber						
8000	3x3	LF	0.023	0.84	2.02		2.86
8020	4x4	"	0.024	1.17	2.15		3.32
8030	Wood						
8040	3x3	LF	0.024	4.36	2.15		6.51

ROUGH CARPENTRY

ID Code	Description Component Descriptions	Output Unit of Meas.	Output Manhr / Unit	Unit Costs Material Cost	Unit Costs Labor Cost	Unit Costs Equipment Cost	Unit Costs Total Cost
06 - 11006	**SLEEPERS**						**06 - 11006**
0960	Sleepers, over concrete						
0980	12" o.c.						
1000	1x2	SF	0.018	0.47	1.61		2.08
1020	1x3	"	0.019	0.69	1.69		2.38
1060	2x4	"	0.022	1.07	1.97		3.04
1080	2x6	"	0.024	1.59	2.08		3.67
1090	16" o.c.						
1100	1x2	SF	0.016	0.35	1.42		1.77
1120	1x3	"	0.016	0.51	1.42		1.93
1140	2x4	"	0.019	0.81	1.69		2.50
1160	2x6	"	0.020	1.19	1.77		2.96
06 - 11007	**SOFFITS**						**06 - 11007**
0980	Soffit framing						
1000	2x3	LF	0.057	0.81	5.07		5.88
1020	2x4	"	0.062	1.07	5.46		6.53
1030	2x6	"	0.067	1.59	5.92		7.51
1040	2x8	"	0.073	2.12	6.45		8.57
06 - 11008	**WALL FRAMING**						**06 - 11008**
0960	Framing wall, studs						
0980	12" o.c.						
1000	2x3	SF	0.015	0.80	1.31		2.11
1040	2x4	"	0.015	1.06	1.31		2.37
1080	2x6	"	0.016	1.59	1.42		3.01
1100	2x8	"	0.017	2.12	1.48		3.60
1110	16" o.c.						
1120	2x3	SF	0.013	0.60	1.11		1.71
1140	2x4	"	0.013	0.80	1.11		1.91
1150	2x6	"	0.013	1.19	1.18		2.37
1160	2x8	"	0.014	1.59	1.22		2.81
1165	24" o.c.						
1170	2x3	SF	0.011	0.40	0.96		1.36
1180	2x4	"	0.011	0.53	0.96		1.49
1190	2x6	"	0.011	0.79	1.01		1.80
1200	2x8	"	0.012	1.06	1.04		2.10
1480	Plates, top or bottom						
1500	2x3	LF	0.024	0.80	2.08		2.88
1510	2x4	"	0.025	1.06	2.22		3.28

ROUGH CARPENTRY

ID Code	Description / Component Descriptions	Output / Unit of Meas.	Output / Manhr / Unit	Unit Costs / Material Cost	Unit Costs / Labor Cost	Unit Costs / Equipment Cost	Total Cost
06 - 11008	**WALL FRAMING, Cont'd...**						**06 - 11008**
1520	2x6	LF	0.027	1.59	2.36		3.95
1530	2x8	"	0.029	2.12	2.53		4.65
2000	Headers, door or window						
2005	2x6						
2008	Single						
2010	3' long	EA	0.400	4.78	35.50		40.25
2020	6' long	"	0.500	9.57	44.50		54.00
2025	Double						
2030	3' long	EA	0.444	9.57	39.50		49.00
2040	6' long	"	0.571	19.25	51.00		70.00
2044	2x8						
2046	Single						
2050	4' long	EA	0.500	8.49	44.50		53.00
2060	8' long	"	0.615	17.00	55.00		72.00
2065	Double						
2070	4' long	EA	0.571	17.00	51.00		68.00
2080	8' long	"	0.727	34.00	65.00		99.00
2085	2x10						
2088	Single						
2090	5' long	EA	0.615	13.25	55.00		68.00
2100	10' long	"	0.800	26.50	71.00		98.00
2110	Double						
2120	5' long	EA	0.667	26.50	59.00		86.00
2130	10' long	"	0.800	53.00	71.00		120
2134	2x12						
2138	Single						
2140	6' long	EA	0.615	19.25	55.00		74.00
2150	12' long	"	0.800	38.25	71.00		110
2155	Double						
2160	6' long	EA	0.727	38.25	65.00		100
2170	12' long	"	0.889	77.00	79.00		160

TIMBER

ID Code	Description Component Descriptions	Output		Unit Costs			
		Unit of Meas.	Manhr / Unit	Material Cost	Labor Cost	Equipment Cost	Total Cost
06 - 13001	**HEAVY TIMBER**						**06 - 13001**
1000	Mill framed structures						
1010	Beams to 20' long						
1020	Douglas fir						
1040	6x8	LF	0.080	13.75	5.56	4.16	23.50
1042	6x10	"	0.083	17.00	5.76	4.31	27.00
1044	6x12	"	0.089	20.50	6.18	4.62	31.25
1046	6x14	"	0.092	23.75	6.42	4.80	35.00
1048	6x16	"	0.096	27.25	6.68	5.00	39.00
1060	8x10	"	0.083	22.75	5.76	4.31	32.75
1070	8x12	"	0.089	27.25	6.18	4.62	38.00
1080	8x14	"	0.092	31.75	6.42	4.80	43.00
1090	8x16	"	0.096	36.25	6.68	5.00	48.00
1200	Southern yellow pine						
1220	6x8	LF	0.080	10.50	5.56	4.16	20.25
1222	6x10	"	0.083	12.75	5.76	4.31	22.75
1224	6x12	"	0.089	16.50	6.18	4.62	27.25
1226	6x14	"	0.092	18.75	6.42	4.80	30.00
1228	6x16	"	0.096	20.75	6.68	5.00	32.50
1240	8x10	"	0.083	17.50	5.76	4.31	27.50
1242	8x12	"	0.089	21.25	6.18	4.62	32.00
1244	8x14	"	0.092	24.25	6.42	4.80	35.50
1246	8x16	"	0.096	28.00	6.68	5.00	39.75
1380	Columns to 12' high						
1400	Douglas fir						
1420	6x6	LF	0.120	10.25	8.35	6.25	24.75
1440	8x8	"	0.120	17.75	8.35	6.25	32.25
1460	10x10	"	0.133	31.00	9.28	6.94	47.25
1480	12x12	"	0.133	38.25	9.28	6.94	55.00
1500	Southern yellow pine						
1520	6x6	LF	0.120	8.25	8.35	6.25	22.75
1540	8x8	"	0.120	13.75	8.35	6.25	28.25
1560	10x10	"	0.133	21.50	9.28	6.94	37.75
1580	12x12	"	0.133	30.00	9.28	6.94	46.25
2000	Posts, treated						
2100	4x4	LF	0.032	2.87	2.84		5.71
2120	6x6	"	0.040	8.33	3.55		12.00

TIMBER

ID Code	Component Descriptions	Unit of Meas.	Manhr / Unit	Material Cost	Labor Cost	Equipment Cost	Total Cost
	Description	**Output**		**Unit Costs**			
06 - 15001	**WOOD DECKING**						**06 - 15001**
0090	Decking, T&G solid						
0095	Cedar						
0100	3" thick	SF	0.020	15.25	1.77		17.00
0120	4" thick	"	0.021	19.00	1.89		21.00
1030	Fir						
1040	3" thick	SF	0.020	6.63	1.77		8.40
1060	4" thick	"	0.021	8.05	1.89		9.94
1080	Southern yellow pine						
2000	3" thick	SF	0.023	6.63	2.02		8.65
2020	4" thick	"	0.025	7.01	2.18		9.19
3120	White pine						
3140	3" thick	SF	0.020	8.05	1.77		9.82
3160	4" thick	"	0.021	11.00	1.89		13.00
5010	Residential decking						
5030	Framing, treated lumber						
5050	12" o.c.						
5070	2x6	SF	0.018	2.20	1.57		3.77
5100	2x8	"	0.018	2.93	1.61		4.54
5150	2x10	"	0.019	3.66	1.65		5.31
5200	2x12	"	0.020	4.40	1.73		6.13
5250	16" o.c.						
5300	2x6	SF	0.015	1.65	1.31		2.96
5350	2x8	"	0.015	2.20	1.34		3.54
5400	2x10	"	0.015	2.75	1.36		4.11
5450	2x12	"	0.016	3.30	1.42		4.72
5500	24" o.c.						
5550	2x6	SF	0.015	1.10	1.34		2.44
5600	2x8	"	0.016	1.47	1.39		2.86
5650	2x10	"	0.017	1.83	1.48		3.31
5700	2x12	"	0.018	2.20	1.57		3.77
5900	Wood Posts, 4x4						
5920	4' high	EA	0.400	11.75	35.50		47.25
5960	6' high	"	0.533	17.50	47.25		65.00
6000	Wood Decking						
6010	Treated, Southern Yellow Pine						
6050	5/4x6	SF	0.064	4.62	5.68		10.25
6100	2x4	"	0.080	4.75	7.10		11.75
6150	2x6	"	0.070	4.62	6.17		10.75

TIMBER

ID Code	Description		Output		Unit Costs			
	Component Descriptions	Unit of Meas.	Manhr / Unit	Material Cost	Labor Cost	Equipment Cost	Total Cost
06 - 15001		**WOOD DECKING, Cont'd...**					**06 - 15001**
6400	Redwood/Cedar						
6450	1x6	SF	0.064	8.25	5.68		14.00
6500	2x4	"	0.080	9.90	7.10		17.00
6550	2x6	"	0.070	9.90	6.17		16.00
8650	Railing, 2x4 rails, 2x2 balusters						
8700	Redwood/cedar	LF	0.160	12.75	14.25		27.00
8750	SYP treated lumber	"	0.160	7.08	14.25		21.25
8800	For composites see Division 06448						

SHEATHING

ID Code							
06 - 16001	**FLOOR SHEATHING**						**06 - 16001**
1980	Sub-flooring, plywood, CDX						
2000	1/2" thick	SF	0.010	0.82	0.88		1.70
2020	5/8" thick	"	0.011	0.93	1.01		1.94
2080	3/4" thick	"	0.013	1.10	1.18		2.28
2090	Structural plywood						
2100	1/2" thick	SF	0.010	1.27	0.88		2.15
2120	5/8" thick	"	0.011	1.44	1.01		2.45
2140	3/4" thick	"	0.012	1.70	1.09		2.79
3100	Board type sub-flooring						
3105	1x6						
3110	Minimum	SF	0.018	1.65	1.57		3.22
3115	Maximum	"	0.020	1.92	1.77		3.69
3117	1x8						
3120	Minimum	SF	0.017	2.20	1.49		3.69
3140	Maximum	"	0.019	2.58	1.67		4.25
3150	1x10						
3160	Minimum	SF	0.016	2.75	1.42		4.17
3180	Maximum	"	0.018	3.24	1.57		4.81
5990	Underlayment						
6000	Hardboard, 1/4" tempered	SF	0.010	1.57	0.88		2.45
6010	Plywood, CDX						
6020	3/8" thick	SF	0.010	0.71	0.88		1.59
6040	1/2" thick	"	0.011	0.82	0.94		1.76
6060	5/8" thick	"	0.011	0.93	1.01		1.94
6080	3/4" thick	"	0.012	1.10	1.09		2.19

SHEATHING

ID Code	Description — Component Descriptions	Output — Unit of Meas.	Output — Manhr / Unit	Unit Costs — Material Cost	Unit Costs — Labor Cost	Unit Costs — Equipment Cost	Unit Costs — Total Cost
06 - 16002	**ROOF SHEATHING**						**06 - 16002**
0080	Sheathing						
0090	Plywood, CDX						
1000	3/8" thick	SF	0.010	0.71	0.91		1.62
1020	1/2" thick	"	0.011	0.82	0.94		1.76
1040	5/8" thick	"	0.011	0.93	1.01		1.94
1060	3/4" thick	"	0.012	1.10	1.09		2.19
1080	Structural plywood						
2040	3/8" thick	SF	0.010	0.44	0.91		1.35
2060	1/2" thick	"	0.011	0.70	0.94		1.64
2080	5/8" thick	"	0.011	0.88	1.01		1.89
2100	3/4" thick	"	0.012	1.04	1.09		2.13
06 - 16003	**WALL SHEATHING**						**06 - 16003**
0980	Sheathing						
0990	Plywood, CDX						
1000	3/8" thick	SF	0.012	0.71	1.05		1.76
1020	1/2" thick	"	0.012	0.82	1.09		1.91
1040	5/8" thick	"	0.013	0.93	1.18		2.11
1060	3/4" thick	"	0.015	1.10	1.29		2.39
3000	Waferboard						
3020	3/8" thick	SF	0.012	0.44	1.05		1.49
3040	1/2" thick	"	0.012	0.70	1.09		1.79
3060	5/8" thick	"	0.013	0.88	1.18		2.06
3080	3/4" thick	"	0.015	1.04	1.29		2.33
4100	Structural plywood						
4120	3/8" thick	SF	0.012	1.10	1.05		2.15
4140	1/2" thick	"	0.012	1.27	1.09		2.36
4160	5/8" thick	"	0.013	1.44	1.18		2.62
4180	3/4" thick	"	0.015	1.70	1.29		2.99
7000	Gypsum, 1/2" thick	"	0.012	0.62	1.09		1.71
7010	5/8" thick	"	0.012	0.94	1.09		2.03
8000	Asphalt impregnated fiberboard, 1/2" thick	"	0.012	1.07	1.09		2.16

TRUSSES

ID Code	Description		Output		Unit Costs			
	Component Descriptions	Unit of Meas.	Manhr / Unit	Material Cost	Labor Cost	Equipment Cost	Total Cost
06 - 17531	**WOOD TRUSSES**						**06 - 17531**
0960	Truss, fink, 2x4 members						
0980	3-in-12 slope						
1000	24' span	EA	0.686	220	47.75	35.75	300
1020	26' span	"	0.686	240	47.75	35.75	320
1021	28' span	"	0.727	260	51.00	37.75	350
1022	30' span	"	0.727	260	51.00	37.75	350
1024	34' span	"	0.774	270	54.00	40.25	360
1025	38' span	"	0.774	280	54.00	40.25	370
1030	5-in-12 slope						
1040	24' span	EA	0.706	240	49.00	36.75	330
1050	28' span	"	0.727	250	51.00	37.75	340
1055	30' span	"	0.750	270	52.00	39.00	360
1060	32' span	"	0.750	290	52.00	39.00	380
1070	40' span	"	0.800	390	56.00	41.75	490
1074	Gable, 2x4 members						
1078	5-in-12 slope						
1080	24' span	EA	0.706	270	49.00	36.75	360
1090	26' span	"	0.706	290	49.00	36.75	380
1100	28' span	"	0.727	330	51.00	37.75	420
1120	30' span	"	0.750	350	52.00	39.00	440
1140	32' span	"	0.750	360	52.00	39.00	450
1160	36' span	"	0.774	380	54.00	40.25	470
1180	40' span	"	0.800	410	56.00	41.75	510
1190	King post type, 2x4 members						
2000	4-in-12 slope						
2040	16' span	EA	0.649	160	45.25	33.75	240
2060	18' span	"	0.667	180	46.50	34.75	260
2080	24' span	"	0.706	190	49.00	36.75	280
2100	26' span	"	0.706	210	49.00	36.75	300
2120	30' span	"	0.750	260	52.00	39.00	350
2140	34' span	"	0.750	280	52.00	39.00	370
2160	38' span	"	0.774	330	54.00	40.25	420
2180	42' span	"	0.828	390	58.00	43.00	490

FINISH CARPENTRY

ID Code	Description / Component Descriptions	Output		Unit Costs			
		Unit of Meas.	Manhr / Unit	Material Cost	Labor Cost	Equipment Cost	Total Cost
06 - 20231	**FINISH CARPENTRY**						**06 - 20231**
0070	Mouldings and trim						
0980	Apron, flat						
1000	9/16 x 2	LF	0.040	1.97	3.55		5.52
1010	9/16 x 3-1/2	"	0.042	4.55	3.73		8.28
1015	Base						
1020	Colonial						
1022	7/16 x 2-1/4	LF	0.040	2.58	3.55		6.13
1024	7/16 x 3	"	0.040	3.34	3.55		6.89
1026	7/16 x 3-1/4	"	0.040	3.42	3.55		6.97
1028	9/16 x 3	"	0.042	3.34	3.73		7.07
1030	9/16 x 3-1/4	"	0.042	3.50	3.73		7.23
1034	11/16 x 2-1/4	"	0.044	3.67	3.94		7.61
1035	Ranch						
1036	7/16 x 2-1/4	LF	0.040	2.83	3.55		6.38
1038	7/16 x 3-1/4	"	0.040	3.34	3.55		6.89
1039	9/16 x 2-1/4	"	0.042	3.08	3.73		6.81
1041	9/16 x 3	"	0.042	3.34	3.73		7.07
1043	9/16 x 3-1/4	"	0.042	3.42	3.73		7.15
1050	Casing						
1060	11/16 x 2-1/2	LF	0.036	2.66	3.22		5.88
1070	11/16 x 3-1/2	"	0.038	3.00	3.38		6.38
1180	Chair rail						
1200	9/16 x 2-1/2	LF	0.040	2.83	3.55		6.38
1210	9/16 x 3-1/2	"	0.040	3.92	3.55		7.47
1250	Closet pole						
1300	1-1/8" dia.	LF	0.053	1.92	4.73		6.65
1310	1-5/8" dia.	"	0.053	2.83	4.73		7.56
1340	Cove						
1500	9/16 x 1-3/4	LF	0.040	2.17	3.55		5.72
1510	11/16 x 2-3/4	"	0.040	3.34	3.55		6.89
1550	Crown						
1600	9/16 x 1-5/8	LF	0.053	2.57	4.73		7.30
1610	9/16 x 2-5/8	"	0.062	2.80	5.46		8.26
1620	11/16 x 3-5/8	"	0.067	3.04	5.92		8.96
1630	11/16 x 4-1/4	"	0.073	4.55	6.45		11.00
1640	11/16 x 5-1/4	"	0.080	5.08	7.10		12.25
1680	Drip cap						
1700	1-1/16 x 1-5/8	LF	0.040	2.73	3.55		6.28

FINISH CARPENTRY

ID Code	Component Descriptions	Unit of Meas.	Manhr / Unit	Material Cost	Labor Cost	Equipment Cost	Total Cost
	Description	**Output**		**Unit Costs**			
06 - 20231	**FINISH CARPENTRY, Cont'd...**						**06 - 20231**
1780	Glass bead						
1800	3/8 x 3/8	LF	0.050	1.08	4.44		5.52
1820	1/2 x 9/16	"	0.050	1.33	4.44		5.77
1840	5/8 x 5/8	"	0.050	1.42	4.44		5.86
1860	3/4 x 3/4	"	0.050	1.67	4.44		6.11
1880	Half round						
1900	1/2	LF	0.032	1.25	2.84		4.09
1910	5/8	"	0.032	1.67	2.84		4.51
1920	3/4	"	0.032	2.26	2.84		5.10
1980	Lattice						
2000	1/4 x 7/8	LF	0.032	0.91	2.84		3.75
2010	1/4 x 1-1/8	"	0.032	0.98	2.84		3.82
2020	1/4 x 1-3/8	"	0.032	1.05	2.84		3.89
2030	1/4 x 1-3/4	"	0.032	1.18	2.84		4.02
2040	1/4 x 2	"	0.032	1.36	2.84		4.20
2080	Ogee molding						
2100	5/8 x 3/4	LF	0.040	1.36	3.55		4.91
2110	11/16 x 1-1/8	"	0.040	3.19	3.55		6.74
2120	11/16 x 1-3/8	"	0.040	2.50	3.55		6.05
2180	Parting bead						
2200	3/8 x 7/8	LF	0.050	1.52	4.44		5.96
2300	Quarter round						
2301	1/4 x 1/4	LF	0.032	0.53	2.84		3.37
2303	3/8 x 3/8	"	0.032	0.76	2.84		3.60
2305	1/2 x 1/2	"	0.032	0.98	2.84		3.82
2307	11/16 x 11/16	"	0.035	0.98	3.08		4.06
2309	3/4 x 3/4	"	0.035	1.81	3.08		4.89
2311	1-1/16 x 1-1/16	"	0.036	1.43	3.22		4.65
2380	Railings, balusters						
2400	1-1/8 x 1-1/8	LF	0.080	4.86	7.10		12.00
2410	1-1/2 x 1-1/2	"	0.073	5.69	6.45		12.25
2480	Screen moldings						
2500	1/4 x 3/4	LF	0.067	1.21	5.92		7.13
2510	5/8 x 5/16	"	0.067	1.52	5.92		7.44
2580	Shoe						
2600	7/16 x 11/16	LF	0.032	1.51	2.84		4.35
2605	Sash beads						
2610	1/2 x 3/4	LF	0.067	1.74	5.92		7.66

FINISH CARPENTRY

ID Code	Component Descriptions	Unit of Meas.	Manhr / Unit	Material Cost	Labor Cost	Equipment Cost	Total Cost
06 - 20231	**FINISH CARPENTRY, Cont'd...**						**06 - 20231**
2620	1/2 x 7/8	LF	0.067	1.97	5.92		7.89
2630	1/2 x 1-1/8	"	0.073	2.12	6.45		8.57
2640	5/8 x 7/8	"	0.073	2.12	6.45		8.57
2760	Stop						
2780	5/8 x 1-5/8						
2800	Colonial	LF	0.050	1.05	4.44		5.49
2810	Ranch	"	0.050	1.05	4.44		5.49
2880	Stools						
2900	11/16 x 2-1/4	LF	0.089	4.63	7.89		12.50
2910	11/16 x 2-1/2	"	0.089	4.86	7.89		12.75
2920	11/16 x 5-1/4	"	0.100	5.01	8.88		14.00
3000	Exterior trim, casing, select pine, 1x3	"	0.040	1.65	3.55		5.20
3040	1x4	"	0.040	2.42	3.55		5.97
3060	1x6	"	0.044	3.85	3.94		7.79
3100	1x8	"	0.050	5.39	4.44		9.83
4010	Douglas fir						
4020	1x3	LF	0.040	1.59	3.55		5.14
4040	1x4	"	0.040	1.96	3.55		5.51
4060	1x6	"	0.044	2.58	3.94		6.52
4100	1x8	"	0.050	3.56	4.44		8.00
4510	Cellular PVC trim						
4520	1x3	LF	0.040	2.33	3.55		5.88
4540	1x4	"	0.040	3.43	3.55		6.98
4560	1x6	"	0.044	4.67	3.94		8.61
4580	1x8	"	0.050	6.18	4.44		10.50
5000	Cornices, white pine, #2 or better						
5020	1x2	LF	0.040	0.99	3.55		4.54
5040	1x4	"	0.040	1.21	3.55		4.76
5060	1x6	"	0.044	1.96	3.94		5.90
5080	1x8	"	0.047	2.42	4.17		6.59
5100	1x10	"	0.050	3.11	4.44		7.55
5120	1x12	"	0.053	3.87	4.73		8.60
6000	Cellular PVC trim						
6020	1x2	LF	0.040	1.71	3.55		5.26
6040	1x4	"	0.040	3.43	3.55		6.98
6060	1x6	"	0.044	4.67	3.94		8.61
6080	1x8	"	0.047	6.18	4.17		10.25
6100	1x10	"	0.050	7.70	4.44		12.25

FINISH CARPENTRY

ID Code	Component Descriptions	Unit of Meas.	Manhr / Unit	Material Cost	Labor Cost	Equipment Cost	Total Cost
	Description	**Output**		**Unit Costs**			
06 - 20231	**FINISH CARPENTRY, Cont'd...**					**06 - 20231**	
6120	1x12	LF	0.053	9.28	4.73		14.00
8600	Shelving, pine						
8620	1x8	LF	0.062	2.62	5.46		8.08
8640	1x10	"	0.064	3.42	5.68		9.10
8660	1x12	"	0.067	4.33	5.92		10.25
8800	Plywood shelf, 3/4", with edge band, 12" wide	"	0.080	4.67	7.10		11.75
8840	Adjustable shelf, and rod, 12" wide						
8860	3' to 4' long	EA	0.200	31.00	17.75		48.75
8880	5' to 8' long	"	0.267	58.00	23.75		82.00
8900	Prefinished wood shelves with brackets and supports						
8905	8" wide						
8910	3' long	EA	0.200	91.00	17.75		110
8922	4' long	"	0.200	100	17.75		120
8924	6' long	"	0.200	150	17.75		170
8930	10" wide						
8940	3' long	EA	0.200	100	17.75		120
8942	4' long	"	0.200	150	17.75		170
8946	6' long	"	0.200	160	17.75		180

MILLWORK

ID Code	Component Descriptions	Unit of Meas.	Manhr / Unit	Material Cost	Labor Cost	Equipment Cost	Total Cost
06 - 22001	**MILLWORK**					**06 - 22001**	
0070	Countertop, laminated plastic						
0080	25" x 7/8" thick						
0099	Minimum	LF	0.200	18.50	17.75		36.25
0100	Average	"	0.267	34.75	23.75		59.00
0110	Maximum	"	0.320	51.00	28.50		80.00
0115	25" x 1-1/4" thick						
0120	Minimum	LF	0.267	22.25	23.75		46.00
0130	Average	"	0.320	44.50	28.50		73.00
0140	Maximum	"	0.400	67.00	35.50		100
0160	Add for cutouts	EA	0.500		44.50		44.50
0165	Backsplash, 4" high, 7/8" thick	LF	0.160	24.50	14.25		38.75
2000	Plywood, sanded, A-C						
2020	1/4" thick	SF	0.027	1.58	2.36		3.94
2040	3/8" thick	"	0.029	1.71	2.53		4.24
2060	1/2" thick	"	0.031	1.94	2.73		4.67
2070	A-D						

MILLWORK

ID Code	Component Descriptions	Unit of Meas.	Manhr / Unit	Material Cost	Labor Cost	Equipment Cost	Total Cost
	Description	**Output**		**Unit Costs**			

ID Code	Component Descriptions	Unit of Meas.	Manhr / Unit	Material Cost	Labor Cost	Equipment Cost	Total Cost
06 - 22001	**MILLWORK, Cont'd...**						**06 - 22001**
2080	1/4" thick	SF	0.027	1.50	2.36		3.86
2090	3/8" thick	"	0.029	1.71	2.53		4.24
2100	1/2" thick	"	0.031	1.86	2.73		4.59
2500	Base cabinet, 34-1/2" high, 24" deep, hardwood						
2540	Minimum	LF	0.320	360	28.50		390
2560	Average	"	0.400	410	35.50		450
2580	Maximum	"	0.533	460	47.25		510
2600	Wall cabinets						
2640	Minimum	LF	0.267	110	23.75		130
2660	Average	"	0.320	150	28.50		180
2680	Maximum	"	0.400	190	35.50		230

ARCHITECTURAL WOODWORK

ID Code	Component Descriptions	Unit of Meas.	Manhr / Unit	Material Cost	Labor Cost	Equipment Cost	Total Cost
06 - 26001	**PANEL WORK**						**06 - 26001**
1020	Hardboard, tempered, 1/4" thick						
1040	Natural faced	SF	0.020	1.63	1.77		3.40
1060	Plastic faced	"	0.023	2.44	2.02		4.46
1080	Pegboard, natural	"	0.020	2.05	1.77		3.82
1100	Plastic faced	"	0.023	1.63	2.02		3.65
1200	Untempered, 1/4" thick						
1220	Natural faced	SF	0.020	1.37	1.77		3.14
1240	Plastic faced	"	0.023	2.37	2.02		4.39
1260	Pegboard, natural	"	0.020	1.46	1.77		3.23
1280	Plastic faced	"	0.023	2.10	2.02		4.12
1300	Plywood, unfinished, 1/4" thick						
1320	Birch						
1330	Natural	SF	0.027	1.83	2.36		4.19
1340	Select	"	0.027	2.69	2.36		5.05
1400	Knotty pine	"	0.027	3.55	2.36		5.91
1500	Cedar (closet lining)						
1520	Standard boards T&G	SF	0.027	4.41	2.36		6.77
1540	Particle board	"	0.027	2.69	2.36		5.05
2000	Plywood, prefinished, 1/4" thick, premium grade						
2020	Birch veneer	SF	0.032	5.16	2.84		8.00
2040	Cherry veneer	"	0.032	6.02	2.84		8.86
2060	Chestnut veneer	"	0.032	7.23	2.84		10.00
2080	Lauan veneer	"	0.032	2.23	2.84		5.07

ARCHITECTURAL WOODWORK

ID Code	Component Descriptions	Unit of Meas.	Manhr / Unit	Material Cost	Labor Cost	Equipment Cost	Total Cost
	Description		**Output**		**Unit Costs**		
06 - 26001	**PANEL WORK, Cont'd...**						**06 - 26001**
2100	Mahogany veneer	SF	0.032	5.94	2.84		8.78
2120	Oak veneer (red)	"	0.032	5.94	2.84		8.78
2140	Pecan veneer	"	0.032	7.49	2.84		10.25
2160	Rosewood veneer	"	0.032	5.03	2.84		7.87
2180	Teak veneer	"	0.032	7.67	2.84		10.50
2200	Walnut veneer	"	0.032	6.63	2.84		9.47
06 - 43131	**STAIRWORK**						**06 - 43131**
0080	Risers, 1x8, 42" wide						
0100	White oak	EA	0.400	46.50	35.50		82.00
0120	Pine	"	0.400	41.25	35.50		77.00
0130	Treads, 1-1/16" x 9-1/2" x 42"						
0140	White oak	EA	0.500	56.00	44.50		100
06 - 44001	**COLUMNS**						**06 - 44001**
0980	Column, hollow, round wood						
0990	12" diameter						
1000	10' high	EA	0.800	1,380	55.00	53.00	1,490
1040	12' high	"	0.857	1,690	59.00	56.00	1,810
1060	14' high	"	0.960	2,030	66.00	63.00	2,160
1080	16' high	"	1.200	2,510	83.00	79.00	2,670
2000	24" diameter						
2020	16' high	EA	1.200	5,740	83.00	79.00	5,900
2040	18' high	"	1.263	6,530	87.00	83.00	6,700
2060	20' high	"	1.263	8,020	87.00	83.00	8,190
2080	22' high	"	1.333	8,450	92.00	88.00	8,630
2100	24' high	"	1.333	9,220	92.00	88.00	9,400
06 - 44800	**PORCH, POST AND RAILINGS**						**06 - 44800**
06 - 44801	**COMPOSITE RESIDENTIAL DECKS**						**06 - 44801**
0100	Composite Decking						
0140	1x6 board						
0160	Minimum	SF	0.067	5.50	5.92		11.50
0180	Average	"	0.080	6.60	7.10		13.75
0200	Maximum	"	0.100	8.25	8.88		17.25
1100	Post						
1150	4x4	EA	0.400	49.50	35.50		85.00
1200	6x6	"	0.533	83.00	47.25		130

ARCHITECTURAL WOODWORK

ID Code	Component Descriptions	Unit of Meas.	Manhr / Unit	Material Cost	Labor Cost	Equipment Cost	Total Cost
	Description	**Output**		**Unit Costs**			
06 - 44801	**COMPOSITE RESIDENTIAL DECKS, Cont'd...**						**06 - 44801**
2000	Post fixtures						
2100	Shoe	EA	0.080	5.22	7.10		12.25
2200	Cap	"	0.080	7.70	7.10		14.75
4000	Railing, including rails and balusters						
4020	Minimum	LF	0.133	44.00	11.75		56.00
4060	Average	"	0.160	55.00	14.25		69.00
4080	Maximum	"	0.200	66.00	17.75		84.00

DIVISION 07
THERMAL AND
MOISTURE

MOISTURE PROTECTION

ID Code	Component Descriptions	Unit of Meas.	Manhr / Unit	Material Cost	Labor Cost	Equipment Cost	Total Cost
	Description	**Output**		**Unit Costs**			

07 - 11001	**DAMPPROOFING**						**07 - 11001**
1000	Silicone dampproofing, sprayed on						
1020	Concrete surface						
1040	1 coat	SF	0.004	0.68	0.30		0.98
1060	2 coats	"	0.006	1.12	0.42		1.54
1070	Concrete block						
1080	1 coat	SF	0.005	0.68	0.37		1.05
1100	2 coats	"	0.007	1.12	0.50		1.62
1110	Brick						
1120	1 coat	SF	0.006	0.78	0.42		1.20
1140	2 coats	"	0.008	1.21	0.55		1.76

07 - 11131	**BITUMINOUS DAMPPROOFING**						**07 - 11131**
0100	Building paper, asphalt felt						
0120	15 lb	SF	0.032	0.22	2.22		2.44
0140	30 lb	"	0.033	0.40	2.32		2.72
1000	Asphalt, troweled, cold, primer plus						
1020	1 coat	SF	0.027	0.74	1.85		2.59
1040	2 coats	"	0.040	1.57	2.78		4.35
1060	3 coats	"	0.050	2.25	3.48		5.73
1200	Fibrous asphalt, hot troweled, primer plus						
1220	1 coat	SF	0.032	0.74	2.22		2.96
1240	2 coats	"	0.044	1.57	3.09		4.66
1260	3 coats	"	0.057	2.25	3.97		6.22
1400	Asphaltic paint dampproofing, per coat						
1420	Brush on	SF	0.011	0.38	0.79		1.17
1440	Spray on	"	0.009	0.55	0.61		1.16

07 - 11161	**PARGING / MASONRY PLASTER**						**07 - 11161**
0080	Parging						
0100	1/2" thick	SF	0.053	0.29	4.52		4.81
0200	3/4" thick	"	0.067	0.37	5.65		6.02
0300	1" thick	"	0.080	0.50	6.78		7.28

07 - 13001	**WATERPROOFING**						**07 - 13001**
1000	Membrane waterproofing, elastomeric						
1020	Butyl						
1040	1/32" thick	SF	0.032	2.01	2.22		4.23
1060	1/16" thick	"	0.033	2.61	2.32		4.93
1140	Neoprene						

MOISTURE PROTECTION

ID Code	Description — Component Descriptions	Output — Unit of Meas.	Output — Manhr / Unit	Unit Costs — Material Cost	Unit Costs — Labor Cost	Unit Costs — Equipment Cost	Unit Costs — Total Cost
07 - 13001	**WATERPROOFING, Cont'd...**						**07 - 13001**
1160	1/32" thick	SF	0.032	3.30	2.22		5.52
1180	1/16" thick	"	0.033	4.73	2.32		7.05
1260	Plastic vapor barrier (polyethylene)						
1280	4 mil	SF	0.003	0.05	0.22		0.27
1300	6 mil	"	0.003	0.08	0.22		0.30
1320	10 mil	"	0.004	0.13	0.27		0.40
1420	Bituminous membrane, asphalt felt, 15 lb.						
1440	One ply	SF	0.020	0.91	1.39		2.30
1460	Two ply	"	0.024	1.07	1.68		2.75
1480	Three ply	"	0.029	1.34	1.98		3.32
2300	Bentonite waterproofing, panels						
2320	3/16" thick	SF	0.020	2.25	1.39		3.64
2330	1/4" thick	"	0.020	2.56	1.39		3.95

INSULATION

ID Code	Description — Component Descriptions	Output — Unit of Meas.	Output — Manhr / Unit	Unit Costs — Material Cost	Unit Costs — Labor Cost	Unit Costs — Equipment Cost	Unit Costs — Total Cost
07 - 21131	**BOARD INSULATION**						**07 - 21131**
2200	Perlite board, roof						
2220	1.00" thick, R2.78	SF	0.007	0.67	0.46		1.13
2240	1.50" thick, R4.17	"	0.007	1.04	0.48		1.52
2580	Rigid urethane						
2600	1" thick, R6.67	SF	0.007	1.15	0.46		1.61
2640	1.50" thick, R11.11	"	0.007	1.56	0.48		2.04
2780	Polystyrene						
2800	1.0" thick, R4.17	SF	0.007	0.77	0.46		1.23
2820	1.5" thick, R6.26	"	0.007	1.17	0.48		1.65
07 - 21161	**BATT INSULATION**						**07 - 21161**
0980	Ceiling, fiberglass, unfaced						
1000	3-1/2" thick, R11	SF	0.009	0.45	0.65		1.10
1020	6" thick, R19	"	0.011	0.59	0.74		1.33
1030	9" thick, R30	"	0.012	1.15	0.85		2.00
1035	Suspended ceiling, unfaced						
1040	3-1/2" thick, R11	SF	0.009	0.49	0.61		1.10
1060	6" thick, R19	"	0.010	0.65	0.69		1.34
1070	9" thick, R30	"	0.011	1.27	0.79		2.06
1075	Crawl space, unfaced						
1080	3-1/2" thick, R11	SF	0.012	0.49	0.85		1.34
1100	6" thick, R19	"	0.013	0.65	0.92		1.57

INSULATION

ID Code	Description — Component Descriptions	Output — Unit of Meas.	Output — Manhr / Unit	Unit Costs — Material Cost	Unit Costs — Labor Cost	Unit Costs — Equipment Cost	Unit Costs — Total Cost
07 - 21161	**BATT INSULATION, Cont'd...**						**07 - 21161**
1120	9" thick, R30	SF	0.015	1.27	1.01		2.28
2000	Wall, fiberglass						
2010	Paper backed						
2020	2" thick, R7	SF	0.008	0.36	0.58		0.94
2040	3" thick, R11	"	0.009	0.40	0.61		1.01
2060	4" thick, R13	"	0.009	0.65	0.65		1.30
2080	6" thick, R19	"	0.010	0.97	0.69		1.66
2082	9" thick, R30	"	0.010	1.76	0.69		2.45
2084	12" thick, R38	"	0.010	2.20	0.69		2.89
2090	Foil backed, 1 side						
2100	2" thick, R8	SF	0.008	0.70	0.58		1.28
2120	3" thick, R11	"	0.009	0.75	0.61		1.36
2140	4" thick, R13	"	0.009	0.78	0.65		1.43
2160	6" thick, R19	"	0.010	1.03	0.69		1.72
2162	9" thick, R30	"	0.010	1.96	0.69		2.65
2164	12" thick, R38	"	0.010	2.31	0.69		3.00
2170	FSK backed 2 sides						
2180	2" thick, R8	SF	0.009	0.80	0.65		1.45
2200	3" thick, R11	"	0.010	1.01	0.69		1.70
2220	4" thick, R13	"	0.011	1.20	0.74		1.94
2240	6" thick, R19	"	0.011	1.29	0.79		2.08
2242	9" thick, R30	"	0.010	2.58	0.69		3.27
2244	12" thick, R38	"	0.010	3.30	0.69		3.99
2250	Unfaced						
2260	2" thick, R8	SF	0.008	0.44	0.58		1.02
2280	3" thick, R11	"	0.009	0.50	0.61		1.11
2300	4" thick, R13	"	0.009	0.54	0.65		1.19
2320	6" thick, R19	"	0.010	0.70	0.69		1.39
2322	9" thick, R30	"	0.010	1.32	0.69		2.01
2324	12" thick, R38	"	0.010	1.96	0.69		2.65
2400	Mineral wool batts						
2410	Paper backed						
2420	2" thick, R6	SF	0.008	0.48	0.58		1.06
2440	4" thick, R12	"	0.009	1.09	0.61		1.70
2460	6" thick, R19	"	0.010	1.36	0.69		2.05
2600	Unfaced						
2610	1" thick, R3	SF	0.008	0.27	0.58		0.85
2620	1-1/2" thick, R4	"	0.009	0.38	0.63		1.01

INSULATION

ID Code	Description		Output		Unit Costs			
	Component Descriptions		Unit of Meas.	Manhr / Unit	Material Cost	Labor Cost	Equipment Cost	Total Cost
07 - 21161			**BATT INSULATION, Cont'd...**					**07 - 21161**
2630	2" thick, R6		SF	0.009	0.48	0.65		1.13
2640	2-1/2" thick, R7		"	0.009	0.61	0.65		1.26
2650	3" thick, R9		"	0.010	0.75	0.67		1.42
2660	3-1/2" thick, R11		"	0.010	0.92	0.67		1.59
2670	4" thick, R12		"	0.010	1.08	0.69		1.77
2680	6" thick, R19		"	0.011	1.36	0.74		2.10
3000	Rigid Safing, flame resistant, unfaced							
3020	1" thick, R4		SF	0.009	1.29	0.65		1.94
3040	2" thick, R8		"	0.010	1.36	0.69		2.05
3060	4" thick, R16		"	0.011	1.76	0.74		2.50
8980	Fasteners, self adhering, attached to ceiling deck							
9000	2-1/2" long		EA	0.013	0.27	0.92		1.19
9020	4-1/2" long		"	0.015	0.30	1.01		1.31
9060	Capped, self-locking washers		"	0.008	0.27	0.55		0.82
07 - 21231			**LOOSE FILL INSULATION**					**07 - 21231**
1000	Blown-in type							
1010	Fiberglass							
1020	5" thick, R11		SF	0.007	0.41	0.46		0.87
1040	6" thick, R13		"	0.008	0.48	0.55		1.03
1060	9" thick, R19		"	0.011	0.58	0.79		1.37
2000	Rockwool, attic application							
2040	6" thick, R13		SF	0.008	0.38	0.55		0.93
2060	8" thick, R19		"	0.010	0.45	0.69		1.14
2080	10" thick, R22		"	0.012	0.53	0.85		1.38
2100	12" thick, R26		"	0.013	0.68	0.92		1.60
2120	15" thick, R30		"	0.016	0.82	1.11		1.93
6200	Poured type							
6210	Fiberglass							
6220	1" thick, R4		SF	0.005	0.45	0.34		0.79
6222	2" thick, R8		"	0.006	0.84	0.39		1.23
6224	3" thick, R12		"	0.007	1.24	0.46		1.70
6226	4" thick, R16		"	0.008	1.63	0.55		2.18
6230	Mineral wool							
6240	1" thick, R3		SF	0.005	0.55	0.34		0.89
6242	2" thick, R6		"	0.006	1.01	0.39		1.40
6244	3" thick, R9		"	0.007	1.54	0.46		2.00
6246	4" thick, R12		"	0.008	1.80	0.55		2.35

INSULATION

ID Code	Component Descriptions	Unit of Meas.	Manhr / Unit	Material Cost	Labor Cost	Equipment Cost	Total Cost
	Description	**Output**		**Unit Costs**			
07 - 21231	**LOOSE FILL INSULATION, Cont'd...**						**07 - 21231**
6300	Vermiculite or perlite						
6310	2" thick, R4.8	SF	0.006	0.98	0.39		1.37
6320	3" thick, R7.2	"	0.007	1.39	0.46		1.85
6330	4" thick, R9.6	"	0.008	1.81	0.55		2.36
8000	Masonry, poured vermiculite or perlite						
8020	4" block	SF	0.004	0.59	0.27		0.86
8040	6" block	"	0.005	0.90	0.34		1.24
8060	8" block	"	0.006	1.31	0.39		1.70
8100	10" block	"	0.006	1.73	0.42		2.15
8120	12" block	"	0.007	2.16	0.46		2.62
07 - 21291	**SPRAYED INSULATION**						**07 - 21291**
1000	Foam, sprayed on						
1010	Polystyrene						
1020	1" thick, R4	SF	0.008	0.91	0.55		1.46
1040	2" thick, R8	"	0.011	1.78	0.74		2.52
1050	Urethane						
1060	1" thick, R4	SF	0.008	0.78	0.55		1.33
1080	2" thick, R8	"	0.011	1.50	0.74		2.24
07 - 26001	**VAPOR BARRIERS**						**07 - 26001**
0980	Vapor barrier, polyethylene						
1000	2 mil	SF	0.004	0.02	0.27		0.29
1010	6 mil	"	0.004	0.08	0.27		0.35
1020	8 mil	"	0.004	0.09	0.30		0.39
1040	10 mil	"	0.004	0.11	0.30		0.41

SHINGLES AND TILES

ID Code	Component Descriptions	Unit of Meas.	Manhr / Unit	Material Cost	Labor Cost	Equipment Cost	Total Cost
07 - 31131	**ASPHALT SHINGLES**						**07 - 31131**
1000	Standard asphalt shingles, strip shingles						
1020	210 lb/square	SQ	0.800	130	68.00		200
1040	235 lb/square	"	0.889	140	75.00		210
1060	240 lb/square	"	1.000	140	85.00		230
1080	260 lb/square	"	1.143	200	97.00		300
1100	300 lb/square	"	1.333	220	110		330
1120	385 lb/square	"	1.600	310	140		450
5980	Roll roofing, mineral surface						
6000	90 lb	SQ	0.571	58.00	48.50		110

SHINGLES AND TILES

ID Code	Component Descriptions	Unit of Meas.	Manhr / Unit	Material Cost	Labor Cost	Equipment Cost	Total Cost
	Description	**Output**		**Unit Costs**			
07 - 31131	**ASPHALT SHINGLES, Cont'd...**						**07 - 31131**
6020	110 lb	SQ	0.667	96.00	57.00		150
6040	140 lb	"	0.800	99.00	68.00		170
7000	Roofing, accessories						
7020	8" Starter strip (roll)	LF	0.020	1.15	1.69		2.84
7040	Starter shingle	"	0.023	0.55	1.93		2.48
7060	Ice & water shield	SF	0.027	0.99	2.26		3.25
07 - 31161	**METAL SHINGLES**						**07 - 31161**
0980	Aluminum, .020" thick						
1000	Plain	SQ	1.600	290	140		430
1020	Colors	"	1.600	320	140		460
1960	Steel, galvanized						
1980	26 ga.						
2000	Plain	SQ	1.600	360	140		500
2020	Colors	"	1.600	460	140		600
2030	24 ga.						
2040	Plain	SQ	1.600	420	140		560
2060	Colors	"	1.600	530	140		670
2960	Porcelain enamel, 22 ga.						
3000	Minimum	SQ	2.000	870	170		1,040
3020	Average	"	2.000	1,000	170		1,170
3040	Maximum	"	2.000	1,120	170		1,290
07 - 31261	**SLATE SHINGLES**						**07 - 31261**
0960	Slate shingles						
0980	Pennsylvania						
1000	Ribbon	SQ	4.000	600	340		940
1020	Clear	"	4.000	770	340		1,110
1030	Vermont						
1040	Black	SQ	4.000	710	340		1,050
1060	Gray	"	4.000	780	340		1,120
1070	Green	"	4.000	800	340		1,140
1080	Red	"	4.000	1,440	340		1,780

SHINGLES AND TILES

ID Code	Component Descriptions	Unit of Meas.	Manhr / Unit	Material Cost	Labor Cost	Equipment Cost	Total Cost
07 - 31291	**WOOD SHINGLES**						**07 - 31291**
1000	Wood shingles, on roofs						
1010	White cedar, #1 shingles						
1020	4" exposure	SQ	2.667	270	230		500
1040	5" exposure	"	2.000	240	170		410
1050	#2 shingles						
1060	4" exposure	SQ	2.667	190	230		420
1080	5" exposure	"	2.000	160	170		330
1090	Resquared and rebutted						
1100	4" exposure	SQ	2.667	240	230		470
1120	5" exposure	"	2.000	200	170		370
1140	On walls						
1150	White cedar, #1 shingles						
1160	4" exposure	SQ	4.000	270	340		610
1180	5" exposure	"	3.200	240	270		510
1200	6" exposure	"	2.667	200	230		430
1210	#2 shingles						
1220	4" exposure	SQ	4.000	190	340		530
1240	5" exposure	"	3.200	160	270		430
1260	6" exposure	"	2.667	130	230		360
1300	Add for fire retarding	"					120
07 - 31292	**WOOD SHAKES**						**07 - 31292**
2010	Shakes, hand split, 24" red cedar, on roofs						
2020	5" exposure	SQ	4.000	800	340		1,140
2040	7" exposure	"	3.200	750	270		1,020
2060	9" exposure	"	2.667	680	230		910
2080	On walls						
2100	6" exposure	SQ	4.000	750	340		1,090
2120	8" exposure	"	3.200	710	270		980
2140	10" exposure	"	2.667	680	230		910
3000	Add for fire retarding	"					120

WALL PANELS

ID Code	Description Component Descriptions	Output		Unit Costs			
		Unit of Meas.	Manhr / Unit	Material Cost	Labor Cost	Equipment Cost	Total Cost
07 - 42005	**METAL SIDING PANELS**						**07 - 42005**
1000	Aluminum siding panels						
1020	Corrugated						
1030	Plain finish						
1040	.024"	SF	0.032	2.23	2.87		5.10
1060	.032"	"	0.032	2.62	2.87		5.49
1070	Painted finish						
1080	.024"	SF	0.032	2.78	2.87		5.65
1100	.032"	"	0.032	3.19	2.87		6.06
2000	Steel siding panels						
2040	Corrugated						
2080	22 ga.	SF	0.053	2.73	4.79		7.52
2100	24 ga.	"	0.053	2.49	4.79		7.28
3000	Ribbed, sheets, galvanized						
3020	22 ga.	SF	0.032	3.42	2.87		6.29
3040	24 ga.	"	0.032	3.01	2.87		5.88
3200	Primed						
3220	24 ga.	SF	0.032	3.01	2.87		5.88
3240	26 ga.	"	0.032	2.46	2.87		5.33

SIDING

ID Code	WOOD SIDING	Output		Unit Costs			07 - 46231
07 - 46231							
1000	Beveled siding, cedar						
1010	A grade						
1040	1/2 x 8	SF	0.032	5.00	2.84		7.84
1060	3/4 x 10	"	0.027	6.43	2.36		8.79
1070	Clear						
1080	1/2 x 6	SF	0.040	5.44	3.55		8.99
1100	1/2 x 8	"	0.032	5.56	2.84		8.40
1120	3/4 x 10	"	0.027	7.45	2.36		9.81
1130	B grade						
1140	1/2 x 6	SF	0.040	5.26	3.55		8.81
1160	1/2 x 8	"	0.032	5.94	2.84		8.78
1180	3/4 x 10	"	0.027	5.60	2.36		7.96
2000	Board and batten						
2010	Cedar						
2020	1x6	SF	0.040	6.84	3.55		10.50
2040	1x8	"	0.032	6.22	2.84		9.06

SIDING

ID Code	Description		Output		Unit Costs			
	Component Descriptions		Unit of Meas.	Manhr / Unit	Material Cost	Labor Cost	Equipment Cost	Total Cost
07 - 46231		**WOOD SIDING, Cont'd...**					**07 - 46231**	
2060	1x10		SF	0.029	5.62	2.53		8.15
2080	1x12		"	0.026	5.04	2.29		7.33
2090	Pine							
2100	1x6		SF	0.040	2.25	3.55		5.80
2120	1x8		"	0.032	2.20	2.84		5.04
2140	1x10		"	0.029	2.11	2.53		4.64
2160	1x12		"	0.026	1.94	2.29		4.23
2170	Redwood							
2180	1x6		SF	0.040	7.44	3.55		11.00
2200	1x8		"	0.032	6.93	2.84		9.77
2220	1x10		"	0.029	6.43	2.53		8.96
2240	1x12		"	0.026	5.93	2.29		8.22
3000	Tongue and groove							
3010	Cedar							
3020	1x4		SF	0.044	7.39	3.94		11.25
3040	1x6		"	0.042	7.11	3.73		10.75
3060	1x8		"	0.040	6.67	3.55		10.25
3080	1x10		"	0.038	6.55	3.38		9.93
3090	Pine							
3100	1x4		SF	0.044	2.44	3.94		6.38
3120	1x6		"	0.042	2.31	3.73		6.04
3140	1x8		"	0.040	2.16	3.55		5.71
3160	1x10		"	0.038	2.05	3.38		5.43
3170	Redwood							
3180	1x4		SF	0.044	6.80	3.94		10.75
3200	1x6		"	0.042	6.55	3.73		10.25
3220	1x8		"	0.040	6.33	3.55		9.88
3240	1x10		"	0.038	6.04	3.38		9.42
07 - 46291		**PLYWOOD SIDING**					**07 - 46291**	
1000	Rough sawn cedar, 3/8" thick		SF	0.027	2.16	2.36		4.52
1020	Fir, 3/8" thick		"	0.027	1.19	2.36		3.55
1980	Texture 1-11, 5/8" thick							
2000	Cedar		SF	0.029	2.92	2.53		5.45
2020	Fir		"	0.029	2.04	2.53		4.57
2040	Redwood		"	0.029	3.14	2.42		5.56
2060	Southern Yellow Pine		"	0.029	1.66	2.53		4.19

SIGN

	Description	Output		Unit Costs			
ID Code	Component Descriptions	Unit of Meas.	Manhr / Unit	Material Cost	Labor Cost	Equipment Cost	Total Cost

SIDING

07 - 46331			PLASTIC SIDING				07 - 46331
1000	Horizontal vinyl siding, solid						
1010	8" wide						
1020	Standard	SF	0.031	1.76	2.73		4.49
1040	Insulated	"	0.031	2.13	2.73		4.86
1050	10" wide						
1060	Standard	SF	0.029	1.82	2.53		4.35
1080	Insulated	"	0.029	2.18	2.53		4.71
8500	Vinyl moldings for doors and windows	LF	0.032	0.87	2.84		3.71

MEMBRANE ROOFING

07 - 51131		BUILT-UP ASPHALT ROOFING					07 - 51131
0980	Built-up roofing, asphalt felt, including gravel						
1000	2 ply	SQ	2.000	120	170		290
1500	3 ply	"	2.667	160	230		390
2000	4 ply	"	3.200	230	270		500
2195	Cant strip, 4" x 4"						
2200	Treated wood	LF	0.023	3.38	1.93		5.31
2260	Foamglass	"	0.020	2.90	1.69		4.59
8000	New gravel for built-up roofing, 400 lb/sq	SQ	1.600	59.00	140		200

07 - 53001		SINGLE-PLY ROOFING					07 - 53001
2000	Elastic sheet roofing						
2060	Neoprene, 1/16" thick	SF	0.010	2.83	0.84		3.67
2115	PVC						
2120	45 mil	SF	0.010	2.34	0.84		3.18
2200	Flashing						
2220	Pipe flashing, 90 mil thick						
2260	1" pipe	EA	0.200	34.00	17.00		51.00
2360	Neoprene flashing, 60 mil thick strip						
2380	6" wide	LF	0.067	2.06	5.65		7.71
2390	12" wide	"	0.100	4.06	8.48		12.50

FLASHING AND SHEET METAL

ID Code	Component Descriptions	Unit of Meas.	Manhr / Unit	Material Cost	Labor Cost	Equipment Cost	Total Cost
	Description	**Output**		**Unit Costs**			
07 - 61001	**METAL ROOFING**						**07 - 61001**
1000	Sheet metal roofing, copper, 16 oz, batten seam	SQ	5.333	1,800	450		2,250
1020	Standing seam	"	5.000	1,760	420		2,180
2000	Aluminum roofing, natural finish						
2005	Corrugated, on steel frame						
2010	.0175" thick	SQ	2.286	140	190		330
2040	.0215" thick	"	2.286	180	190		370
2060	.024" thick	"	2.286	210	190		400
2080	.032" thick	"	2.286	260	190		450
2100	V-beam, on steel frame						
2120	.032" thick	SQ	2.286	270	190		460
2130	.040" thick	"	2.286	290	190		480
2140	.050" thick	"	2.286	370	190		560
2200	Ridge cap						
2220	.019" thick	LF	0.027	4.24	2.26		6.50
2500	Corrugated galvanized steel roofing, on steel frame						
2520	28 ga.	SQ	2.286	230	190		420
2540	26 ga.	"	2.286	270	190		460
2550	24 ga.	"	2.286	300	190		490
2560	22 ga.	"	2.286	330	190		520
2580	26 ga., factory insulated with 1" polystyrene	"	3.200	510	270		780
2600	Ridge roll						
2620	10" wide	LF	0.027	2.31	2.26		4.57
2640	20" wide	"	0.032	4.70	2.71		7.41
07 - 62001	**FLASHING AND TRIM**						**07 - 62001**
0050	Counter flashing						
0060	Aluminum, .032"	SF	0.080	2.51	6.78		9.29
0100	Stainless steel, .015"	"	0.080	8.03	6.78		14.75
0105	Copper						
0110	16 oz.	SF	0.080	7.48	6.78		14.25
0112	20 oz.	"	0.080	8.89	6.78		15.75
0114	24 oz.	"	0.080	10.75	6.78		17.50
0116	32 oz.	"	0.080	13.25	6.78		20.00
0118	Valley flashing						
0120	Aluminum, .032"	SF	0.050	1.74	4.24		5.98
0130	Stainless steel, .015	"	0.050	5.56	4.24		9.80
0135	Copper						
0140	16 oz.	SF	0.050	7.48	4.24		11.75

FLASHING AND SHEET METAL

ID Code	Description	Output		Unit Costs			
	Component Descriptions	Unit of Meas.	Manhr / Unit	Material Cost	Labor Cost	Equipment Cost	Total Cost
07 - 62001	**FLASHING AND TRIM, Cont'd...**						**07 - 62001**
0160	20 oz.	SF	0.067	8.89	5.65		14.50
0180	24 oz.	"	0.050	10.75	4.24		15.00
0200	32 oz.	"	0.050	13.25	4.24		17.50
0380	Base flashing						
0400	Aluminum, .040"	SF	0.067	2.60	5.65		8.25
0410	Stainless steel, .018"	"	0.067	6.65	5.65		12.25
0415	Copper						
0420	16 oz.	SF	0.067	15.00	5.65		20.75
0422	20 oz.	"	0.050	17.75	4.24		22.00
0424	24 oz.	"	0.067	21.50	5.65		27.25
0426	32 oz.	"	0.067	26.50	5.65		32.25
3201	Flashing and trim, aluminum						
3221	.019" thick	SF	0.057	2.25	4.84		7.09
3231	.032" thick	"	0.057	2.75	4.84		7.59
3241	.040" thick	"	0.062	4.71	5.22		9.93
3310	Neoprene sheet flashing, .060" thick	"	0.050	3.75	4.24		7.99
3320	Copper, paper backed						
3330	2 oz.	SF	0.080	4.81	6.78		11.50
3340	5 oz.	"	0.080	6.21	6.78		13.00
07 - 71231	**GUTTERS AND DOWNSPOUTS**						**07 - 71231**
1500	Copper gutter and downspout						
1520	Downspouts, 16 oz. copper						
1530	Round						
1540	3" dia.	LF	0.053	15.50	4.52		20.00
1550	4" dia.	"	0.053	19.50	4.52		24.00
1560	Rectangular, corrugated						
1570	2" x 3"	LF	0.050	15.00	4.24		19.25
1580	3" x 4"	"	0.050	18.50	4.24		22.75
1585	Rectangular, flat surface						
1590	2" x 3"	LF	0.053	17.25	4.52		21.75
1600	3" x 4"	"	0.053	24.50	4.52		29.00
1620	Lead-coated copper downspouts						
1625	Round						
1630	3" dia.	LF	0.050	24.50	4.24		28.75
1650	4" dia.	"	0.057	29.50	4.84		34.25
1670	Rectangular, corrugated						
1680	2" x 3"	LF	0.053	20.50	4.52		25.00

FLASHING AND SHEET METAL

ID Code	Description / Component Descriptions	Output Unit of Meas.	Manhr / Unit	Material Cost	Labor Cost	Equipment Cost	Total Cost
07 - 71231	**GUTTERS AND DOWNSPOUTS, Cont'd...**						**07 - 71231**
1690	3" x 4"	LF	0.053	24.50	4.52		29.00
1695	Rectangular, plain						
1700	2" x 3"	LF	0.053	14.25	4.52		18.75
1750	3" x 4"	"	0.053	16.50	4.52		21.00
1800	Gutters, 16 oz. copper						
1810	Half round						
1820	4" wide	LF	0.080	14.00	6.78		20.75
1840	5" wide	"	0.089	17.00	7.54		24.50
1860	Type K						
1880	4" wide	LF	0.080	15.50	6.78		22.25
1890	5" wide	"	0.089	16.25	7.54		23.75
1900	Lead-coated copper gutters						
1905	Half round						
1910	4" wide	LF	0.080	17.00	6.78		23.75
1920	6" wide	"	0.089	23.25	7.54		30.75
1925	Type K						
1930	4" wide	LF	0.080	18.50	6.78		25.25
1940	5" wide	"	0.089	24.00	7.54		31.50
3000	Aluminum gutter and downspout						
3005	Downspouts						
3010	2" x 3"	LF	0.053	2.19	4.52		6.71
3030	3" x 4"	"	0.057	3.02	4.84		7.86
3035	4" x 5"	"	0.062	3.48	5.22		8.70
3038	Round						
3040	3" dia.	LF	0.053	3.51	4.52		8.03
3050	4" dia.	"	0.057	4.50	4.84		9.34
3240	Gutters, stock units						
3260	4" wide	LF	0.084	2.60	7.14		9.74
3270	5" wide	"	0.089	3.09	7.54		10.75
4101	Galvanized steel gutter and downspout						
4111	Downspouts, round corrugated						
4121	3" dia.	LF	0.053	2.87	4.52		7.39
4131	4" dia.	"	0.053	3.85	4.52		8.37
4141	5" dia.	"	0.057	5.74	4.84		10.50
4151	6" dia.	"	0.057	7.62	4.84		12.50
4161	Rectangular						
4171	2" x 3"	LF	0.053	2.60	4.52		7.12
4191	3" x 4"	"	0.050	3.72	4.24		7.96

FLASHING AND SHEET METAL

ID Code	Description / Component Descriptions	Output / Unit of Meas.	Output / Manhr / Unit	Unit Costs / Material Cost	Unit Costs / Labor Cost	Unit Costs / Equipment Cost	Unit Costs / Total Cost
07 - 71231	**GUTTERS AND DOWNSPOUTS, Cont'd...**						**07 - 71231**
4201	4" x 4"	LF	0.050	4.65	4.24		8.89
4300	Gutters, stock units						
4310	5" wide						
4320	Plain	LF	0.089	2.51	7.54		10.00
4330	Painted	"	0.089	2.72	7.54		10.25
4335	6" wide						
4340	Plain	LF	0.094	3.50	7.98		11.50
4360	Painted	"	0.094	3.93	7.98		12.00

JOINT SEALANTS

ID Code	Component Descriptions	Unit of Meas.	Manhr / Unit	Material Cost	Labor Cost	Equipment Cost	Total Cost
07 - 92001	**CAULKING**						**07 - 92001**
0100	Caulk exterior, two component						
0120	1/4 x 1/2	LF	0.040	0.43	3.55		3.98
0140	3/8 x 1/2	"	0.044	0.66	3.94		4.60
0160	1/2 x 1/2	"	0.050	0.90	4.44		5.34
0220	Caulk interior, single component						
0240	1/4 x 1/2	LF	0.038	0.29	3.38		3.67
0260	3/8 x 1/2	"	0.042	0.41	3.73		4.14
0280	1/2 x 1/2	"	0.047	0.54	4.17		4.71

DIVISION 08
DOORS AND WINDOWS

METAL DOORS & TRANSOMS

ID Code	Description Component Descriptions	Output Unit of Meas.	Manhr / Unit	Unit Costs Material Cost	Labor Cost	Equipment Cost	Total Cost
08 - 11131	**METAL DOORS**						**08 - 11131**
1000	Flush hollow metal, std. duty, 20 ga., 1-3/8" thick						
1020	2-6 x 6-8	EA	0.889	570	79.00		650
1040	2-8 x 6-8	"	0.889	600	79.00		680
1080	3-0 x 6-8	"	0.889	640	79.00		720
1090	1-3/4" thick						
1100	2-6 x 6-8	EA	0.889	580	79.00		660
1120	2-8 x 6-8	"	0.889	710	79.00		790
1150	3-0 x 6-8	"	0.889	660	79.00		740
1200	2-6 x 7-0	"	0.889	640	79.00		720
1210	2-8 x 7-0	"	0.889	660	79.00		740
1240	3-0 x 7-0	"	0.889	700	79.00		780
2110	Heavy duty, 20 ga., unrated, 1-3/4"						
2130	2-8 x 6-8	EA	0.889	640	79.00		720
2135	3-0 x 6-8	"	0.889	680	79.00		760
2140	2-8 x 7-0	"	0.889	720	79.00		800
2150	3-0 x 7-0	"	0.889	700	79.00		780
2170	3-4 x 7-0	"	0.889	720	79.00		800
2200	18 ga., 1-3/4", unrated door						
2210	2-0 x 7-0	EA	0.889	670	79.00		750
2230	2-4 x 7-0	"	0.889	670	79.00		750
2235	2-6 x 7-0	"	0.889	670	79.00		750
2240	2-8 x 7-0	"	0.889	740	79.00		820
2260	3-0 x 7-0	"	0.889	680	79.00		760
2270	3-4 x 7-0	"	0.889	840	79.00		920
2310	2", unrated door						
2320	2-0 x 7-0	EA	1.000	740	89.00		830
2330	2-4 x 7-0	"	1.000	740	89.00		830
2340	2-6 x 7-0	"	1.000	740	89.00		830
2350	2-8 x 7-0	"	1.000	810	89.00		900
2360	3-0 x 7-0	"	1.000	840	89.00		930
2370	3-4 x 7-0	"	1.000	860	89.00		950
08 - 11134	**METAL DOOR FRAMES**						**08 - 11134**
1000	Hollow metal, stock, 18 ga., 4-3/4" x 1-3/4"						
1020	2-0 x 7-0	EA	1.000	160	89.00		250
1040	2-4 x 7-0	"	1.000	180	89.00		270
1060	2-6 x 7-0	"	1.000	180	89.00		270
1080	2-8 x 7-0	"	1.000	180	89.00		270

METAL DOORS & TRANSOMS

ID Code	Description / Component Descriptions	Output Unit of Meas.	Output Manhr / Unit	Unit Costs Material Cost	Labor Cost	Equipment Cost	Total Cost
08 - 11134	**METAL DOOR FRAMES, Cont'd...**						**08 - 11134**
1100	3-0 x 7-0	EA	1.000	180	89.00		270
1120	4-0 x 7-0	"	1.333	200	120		320
1140	5-0 x 7-0	"	1.333	210	120		330
1160	6-0 x 7-0	"	1.333	250	120		370
1500	16 ga., 6-3/4" x 1-3/4"						
1520	2-0 x 7-0	EA	1.000	180	90.00		270
1530	2-4 x 7-0	"	1.000	170	90.00		260
1535	2-6 x 7-0	"	1.000	170	90.00		260
1540	2-8 x 7-0	"	1.000	180	90.00		270
1550	3-0 x 7-0	"	1.000	190	90.00		280
1560	4-0 x 7-0	"	1.333	220	120		340
1580	6-0 x 7-0	"	1.333	260	120		380

WOOD AND PLASTIC

ID Code	Component Descriptions	Unit of Meas.	Manhr / Unit	Material Cost	Labor Cost	Equipment Cost	Total Cost
08 - 14001	**WOOD DOORS**						**08 - 14001**
0980	Solid core, 1-3/8" thick						
1000	Birch faced						
1020	2-4 x 7-0	EA	1.000	190	89.00		280
1040	2-8 x 7-0	"	1.000	200	89.00		290
1060	3-0 x 7-0	"	1.000	200	89.00		290
1070	3-4 x 7-0	"	1.000	400	89.00		490
1080	2-4 x 6-8	"	1.000	190	89.00		280
1090	2-6 x 6-8	"	1.000	190	89.00		280
1095	2-8 x 6-8	"	1.000	200	89.00		290
1100	3-0 x 6-8	"	1.000	200	89.00		290
1120	Lauan faced						
1140	2-4 x 6-8	EA	1.000	160	89.00		250
1160	2-8 x 6-8	"	1.000	170	89.00		260
1180	3-0 x 6-8	"	1.000	180	89.00		270
1200	3-4 x 6-8	"	1.000	290	89.00		380
1300	Tempered hardboard faced						
1320	2-4 x 7-0	EA	1.000	200	89.00		290
1340	2-8 x 7-0	"	1.000	210	89.00		300
1360	3-0 x 7-0	"	1.000	240	89.00		330
1380	3-4 x 7-0	"	1.000	250	89.00		340
1420	Hollow core, 1-3/8" thick						
1440	Birch faced						

WOOD AND PLASTIC

ID Code	Component Descriptions	Unit of Meas.	Manhr / Unit	Material Cost	Labor Cost	Equipment Cost	Total Cost
08 - 14001	**WOOD DOORS, Cont'd...**						**08 - 14001**
1460	2-4 x 7-0	EA	1.000	160	89.00		250
1480	2-8 x 7-0	"	1.000	160	89.00		250
1500	3-0 x 7-0	"	1.000	170	89.00		260
1520	3-4 x 7-0	"	1.000	180	89.00		270
1600	Lauan faced						
1620	2-4 x 6-8	EA	1.000	81.00	89.00		170
1630	2-6 x 6-8	"	1.000	88.00	89.00		180
1640	2-8 x 6-8	"	1.000	110	89.00		200
1660	3-0 x 6-8	"	1.000	110	89.00		200
1680	3-4 x 6-8	"	1.000	130	89.00		220
1740	Tempered hardboard faced						
1760	2-4 x 7-0	EA	1.000	83.00	89.00		170
1770	2-6 x 7-0	"	1.000	89.00	89.00		180
1780	2-8 x 7-0	"	1.000	98.00	89.00		190
1800	3-0 x 7-0	"	1.000	100	89.00		190
1820	3-4 x 7-0	"	1.000	110	89.00		200
1900	Solid core, 1-3/4" thick						
1920	Birch faced						
1940	2-4 x 7-0	EA	1.000	270	89.00		360
1950	2-6 x 7-0	"	1.000	270	89.00		360
1960	2-8 x 7-0	"	1.000	290	89.00		380
1970	3-0 x 7-0	"	1.000	270	89.00		360
1980	3-4 x 7-0	"	1.000	270	89.00		360
2000	Lauan faced						
2020	2-4 x 7-0	EA	1.000	190	89.00		280
2030	2-6 x 7-0	"	1.000	210	89.00		300
2040	2-8 x 7-0	"	1.000	230	89.00		320
2060	3-0 x 7-0	"	1.000	240	89.00		330
2080	3-4 x 7-0	"	1.000	250	89.00		340
2140	Tempered hardboard faced						
2160	2-4 x 7-0	EA	1.000	200	89.00		290
2170	2-6 x 7-0	"	1.000	220	89.00		310
2180	2-8 x 7-0	"	1.000	240	89.00		330
2190	3-0 x 7-0	"	1.000	250	89.00		340
2200	3-4 x 7-0	"	1.000	270	89.00		360
2250	Hollow core, 1-3/4" thick						
2270	Birch faced						
2290	2-4 x 7-0	EA	1.000	190	89.00		280

WOOD AND PLASTIC

ID Code	Description — Component Descriptions	Output — Unit of Meas.	Output — Manhr / Unit	Unit Costs — Material Cost	Unit Costs — Labor Cost	Unit Costs — Equipment Cost	Unit Costs — Total Cost
08 - 14001	**WOOD DOORS, Cont'd...**						**08 - 14001**
2295	2-6 x 7-0	EA	1.000	190	89.00		280
2300	2-8 x 7-0	"	1.000	200	89.00		290
2320	3-0 x 7-0	"	1.000	200	89.00		290
2340	3-4 x 7-0	"	1.000	210	89.00		300
2400	Lauan faced						
2420	2-4 x 6-8	EA	1.000	110	89.00		200
2430	2-6 x 6-8	"	1.000	120	89.00		210
2440	2-8 x 6-8	"	1.000	110	89.00		200
2460	3-0 x 6-8	"	1.000	120	89.00		210
2480	3-4 x 6-8	"	1.000	120	89.00		210
2520	Tempered hardboard						
2540	2-4 x 7-0	EA	1.000	100	89.00		190
2550	2-6 x 7-0	"	1.000	110	89.00		200
2560	2-8 x 7-0	"	1.000	110	89.00		200
2580	3-0 x 7-0	"	1.000	120	89.00		210
2600	3-4 x 7-0	"	1.000	120	89.00		210
2620	Add-on, louver	"	0.800	34.50	71.00		110
2640	Glass	"	0.800	110	71.00		180
2700	Exterior doors, 3-0 x 7-0 x 2-1/2", solid core						
2710	Carved						
2720	One face	EA	2.000	1,150	180		1,330
2740	Two faces	"	2.000	1,590	180		1,770
3000	Closet doors, 1-3/4" thick						
3001	Bi-fold or bi-passing, includes frame and trim						
3020	Paneled						
3040	4-0 x 6-8	EA	1.333	460	120		580
3060	6-0 x 6-8	"	1.333	520	120		640
3070	Louvered						
3080	4-0 x 6-8	EA	1.333	280	120		400
3100	6-0 x 6-8	"	1.333	330	120		450
3130	Flush						
3140	4-0 x 6-8	EA	1.333	200	120		320
3160	6-0 x 6-8	"	1.333	260	120		380
3170	Primed						
3180	4-0 x 6-8	EA	1.333	220	120		340
3200	6-0 x 6-8	"	1.333	250	120		370

WOOD AND PLASTIC

ID Code	Description — Component Descriptions	Output — Unit of Meas.	Output — Manhr / Unit	Unit Costs — Material Cost	Unit Costs — Labor Cost	Unit Costs — Equipment Cost	Unit Costs — Total Cost
08 - 14009	**WOOD FRAMES**						**08 - 14009**
0080	Frame, interior, pine						
0100	2-6 x 6-8	EA	1.143	83.00	100		180
0140	2-8 x 6-8	"	1.143	89.00	100		190
0160	3-0 x 6-8	"	1.143	92.00	100		190
0180	5-0 x 6-8	"	1.143	96.00	100		200
0200	6-0 x 6-8	"	1.143	100	100		200
0220	2-6 x 7-0	"	1.143	95.00	100		200
0240	2-8 x 7-0	"	1.143	110	100		210
0260	3-0 x 7-0	"	1.143	110	100		210
0280	5-0 x 7-0	"	1.600	120	140		260
0300	6-0 x 7-0	"	1.600	130	140		270
1000	Exterior, custom, with threshold, including trim						
1040	Walnut						
1060	3-0 x 7-0	EA	2.000	330	180		510
1080	6-0 x 7-0	"	2.000	380	180		560
1090	Oak						
1100	3-0 x 7-0	EA	2.000	300	180		480
1120	6-0 x 7-0	"	2.000	340	180		520
1200	Pine						
1220	2-4 x 7-0	EA	1.600	120	140		260
1240	2-6 x 7-0	"	1.600	130	140		270
1280	2-8 x 7-0	"	1.600	160	140		300
1300	3-0 x 7-0	"	1.600	170	140		310
1320	3-4 x 7-0	"	1.600	180	140		320
1340	6-0 x 7-0	"	2.667	190	240		430
08 - 34001	**SPECIAL DOORS**						**08 - 34001**
6400	Sliding glass doors						
6410	Tempered plate glass, 1/4" thick						
6420	6' wide						
6440	Economy grade	EA	2.667	1,130	240		1,370
6450	Premium grade	"	2.667	1,820	240		2,060
6455	12' wide						
6460	Economy grade	EA	4.000	1,590	360		1,950
6465	Premium grade	"	4.000	2,390	360		2,750
6470	Insulating glass, 5/8" thick						
6475	6' wide						
6480	Economy grade	EA	2.667	1,400	240		1,640

WOOD AND PLASTIC

ID Code	Description — Component Descriptions	Output — Unit of Meas.	Output — Manhr / Unit	Unit Costs — Material Cost	Unit Costs — Labor Cost	Unit Costs — Equipment Cost	Unit Costs — Total Cost
08 - 34001	**SPECIAL DOORS, Cont'd...**						**08 - 34001**
6490	Premium grade	EA	2.667	1,800	240		2,040
6500	12' wide						
6510	Economy grade	EA	4.000	1,740	360		2,100
6515	Premium grade	"	4.000	2,790	360		3,150
6520	1" thick						
6525	6' wide						
6530	Economy grade	EA	2.667	1,760	240		2,000
6535	Premium grade	"	2.667	2,030	240		2,270
6540	12' wide						
6550	Economy grade	EA	4.000	2,730	360		3,090
6560	Premium grade	"	4.000	3,990	360		4,350
6600	Added costs						
6610	Custom quality, add to material, 30%						
6630	Tempered glass, 6' wide, add	SF					4.90
6850	Residential storm door						
6900	Minimum	EA	1.333	180	120		300
6920	Average	"	1.333	230	120		350
6940	Maximum	"	2.000	510	180		690

GLAZED CURTAIN WALLS

ID Code	Description — Component Descriptions	Output — Unit of Meas.	Output — Manhr / Unit	Unit Costs — Material Cost	Unit Costs — Labor Cost	Unit Costs — Equipment Cost	Unit Costs — Total Cost
08 - 44001	**GLAZED CURTAIN WALLS**						**08 - 44001**
1000	Curtain wall, aluminum system, framing sections						
1005	2" x 3"						
1010	Jamb	LF	0.067	19.75	5.99		25.75
1020	Horizontal	"	0.067	20.00	5.99		26.00
1030	Mullion	"	0.067	26.75	5.99		32.75
1035	2" x 4"						
1040	Jamb	LF	0.100	26.75	8.99		35.75
1060	Horizontal	"	0.100	27.50	8.99		36.50
1070	Mullion	"	0.100	26.75	8.99		35.75
1080	3" x 5-1/2"						
1090	Jamb	LF	0.100	35.50	8.99		44.50
1100	Horizontal	"	0.100	39.50	8.99		48.50
1110	Mullion	"	0.100	35.75	8.99		44.75
1115	4" corner mullion	"	0.133	47.25	12.00		59.00
1120	Coping sections						
1130	1/8" x 8"	LF	0.133	45.00	12.00		57.00

GLAZED CURTAIN WALLS

ID Code	Component Descriptions	Unit of Meas.	Manhr / Unit	Material Cost	Labor Cost	Equipment Cost	Total Cost
	Description	**Output**		**Unit Costs**			

08 - 44001   **GLAZED CURTAIN WALLS, Cont'd...**   **08 - 44001**

ID Code	Component Descriptions	Unit of Meas.	Manhr / Unit	Material Cost	Labor Cost	Equipment Cost	Total Cost
1140	1/8" x 9"	LF	0.133	45.25	12.00		57.00
1150	1/8" x 12-1/2"	"	0.160	46.25	14.50		61.00
1160	Sill section						
1170	1/8" x 6"	LF	0.080	44.25	7.19		51.00
1180	1/8" x 7"	"	0.080	44.75	7.19		52.00
1190	1/8" x 8-1/2"	"	0.080	45.75	7.19		53.00
1200	Column covers, aluminum						
1210	1/8" x 26"	LF	0.200	66.00	18.00		84.00
1220	1/8" x 34"	"	0.211	74.00	19.00		93.00
1230	1/8" x 38"	"	0.211	75.00	19.00		94.00
1500	Doors						
1600	Aluminum framed, standard hardware						
1620	Narrow stile						
1630	2-6 x 7-0	EA	4.000	820	360		1,180
1640	3-0 x 7-0	"	4.000	820	360		1,180
1660	3-6 x 7-0	"	4.000	850	360		1,210

METAL WINDOWS

08 - 51131   **ALUMINUM WINDOWS**   **08 - 51131**

ID Code	Component Descriptions	Unit of Meas.	Manhr / Unit	Material Cost	Labor Cost	Equipment Cost	Total Cost
0110	Jalousie						
0120	3-0 x 4-0	EA	1.000	360	90.00		450
0140	3-0 x 5-0	"	1.000	420	90.00		510
0220	Fixed window						
0240	6 sf to 8 sf	SF	0.114	19.75	10.25		30.00
0250	12 sf to 16 sf	"	0.089	17.75	7.99		25.75
0255	Projecting window						
0260	6 sf to 8 sf	SF	0.200	44.00	18.00		62.00
0270	12 sf to 16 sf	"	0.133	39.75	12.00		52.00
0275	Horizontal sliding						
0280	6 sf to 8 sf	SF	0.100	28.50	8.99		37.50
0290	12 sf to 16 sf	"	0.080	26.50	7.19		33.75
1140	Double hung						
1160	6 sf to 8 sf	SF	0.160	39.75	14.50		54.00
1180	10 sf to 12 sf	"	0.133	35.25	12.00		47.25
3010	Storm window, 0.5 cfm, up to						
3020	60 u.i. (united inches)	EA	0.400	92.00	36.00		130
3040	70 u.i.	"	0.400	95.00	36.00		130

METAL WINDOWS

ID Code	Description		Output		Unit Costs			
		Component Descriptions	Unit of Meas.	Manhr / Unit	Material Cost	Labor Cost	Equipment Cost	Total Cost

08 - 51131 — ALUMINUM WINDOWS, Cont'd... — 08 - 51131

ID Code	Component Descriptions	Unit of Meas.	Manhr / Unit	Material Cost	Labor Cost	Equipment Cost	Total Cost
3060	80 u.i.	EA	0.400	110	36.00		150
3080	90 u.i.	"	0.444	110	40.00		150
3100	100 u.i.	"	0.444	110	40.00		150
3110	2.0 cfm, up to						
3120	60 u.i.	EA	0.400	120	36.00		160
3140	70 u.i.	"	0.400	120	36.00		160
3160	80 u.i.	"	0.400	120	36.00		160
3180	90 u.i.	"	0.444	130	40.00		170
3200	100 u.i.	"	0.444	130	40.00		170

WOOD & PLASTIC

08 - 52001 — WOOD WINDOWS — 08 - 52001

ID Code	Component Descriptions	Unit of Meas.	Manhr / Unit	Material Cost	Labor Cost	Equipment Cost	Total Cost
0980	Double hung						
0990	24" x 36"						
1000	Minimum	EA	0.800	250	71.00		320
1002	Average	"	1.000	370	89.00		460
1004	Maximum	"	1.333	490	120		610
1010	24" x 48"						
1020	Minimum	EA	0.800	300	71.00		370
1022	Average	"	1.000	440	89.00		530
1024	Maximum	"	1.333	620	120		740
1030	30" x 48"						
1040	Minimum	EA	0.889	320	79.00		400
1042	Average	"	1.143	450	100		550
1044	Maximum	"	1.600	640	140		780
1050	30" x 60"						
1060	Minimum	EA	0.889	350	79.00		430
1062	Average	"	1.143	560	100		660
1064	Maximum	"	1.600	680	140		820
1160	Casement						
1180	1 leaf, 22" x 38" high						
1220	Minimum	EA	0.800	380	71.00		450
1222	Average	"	1.000	460	89.00		550
1224	Maximum	"	1.333	540	120		660
1230	2 leaf, 50" x 50" high						
1240	Minimum	EA	1.000	1,020	89.00		1,110
1242	Average	"	1.333	1,330	120		1,450

WOOD & PLASTIC

ID Code	Component Descriptions	Unit of Meas.	Manhr / Unit	Material Cost	Labor Cost	Equipment Cost	Total Cost
	Description	**Output**		**Unit Costs**			
08 - 52001	**WOOD WINDOWS, Cont'd...**						**08 - 52001**
1244	Maximum	EA	2.000	1,520	180		1,700
1250	3 leaf, 71" x 62" high						
1260	Minimum	EA	1.000	1,460	89.00		1,550
1262	Average	"	1.333	1,490	120		1,610
1264	Maximum	"	2.000	1,780	180		1,960
1270	4 leaf, 95" x 75" high						
1280	Minimum	EA	1.143	1,940	100		2,040
1282	Average	"	1.600	2,210	140		2,350
1284	Maximum	"	2.667	2,820	240		3,060
1290	5 leaf, 119" x 75" high						
1300	Minimum	EA	1.143	2,510	100		2,610
1302	Average	"	1.600	2,700	140		2,840
1304	Maximum	"	2.667	3,460	240		3,700
1360	Picture window, fixed glass, 54" x 54" high						
1400	Minimum	EA	1.000	520	89.00		610
1422	Average	"	1.143	580	100		680
1424	Maximum	"	1.333	1,030	120		1,150
1430	68" x 55" high						
1440	Minimum	EA	1.000	930	89.00		1,020
1442	Average	"	1.143	1,070	100		1,170
1444	Maximum	"	1.333	1,400	120		1,520
1480	Sliding, 40" x 31" high						
1520	Minimum	EA	0.800	310	71.00		380
1522	Average	"	1.000	470	89.00		560
1524	Maximum	"	1.333	560	120		680
1530	52" x 39" high						
1540	Minimum	EA	1.000	380	89.00		470
1542	Average	"	1.143	570	100		670
1544	Maximum	"	1.333	610	120		730
1550	64" x 72" high						
1560	Minimum	EA	1.000	600	89.00		690
1562	Average	"	1.333	950	120		1,070
1564	Maximum	"	1.600	1,050	140		1,190
1760	Awning windows						
1780	34" x 21" high						
1800	Minimum	EA	0.800	310	71.00		380
1822	Average	"	1.000	360	89.00		450
1824	Maximum	"	1.333	420	120		540

WOOD & PLASTIC

ID Code	Description		Output		Unit Costs			
	Component Descriptions	Unit of Meas.	Manhr / Unit	Material Cost	Labor Cost	Equipment Cost	Total Cost
08 - 52001	**WOOD WINDOWS, Cont'd...**						**08 - 52001**	
1840	40" x 21" high						
1860	Minimum	EA	0.889	360	79.00		440
1862	Average	"	1.143	410	100		510
1864	Maximum	"	1.600	450	140		590
1880	48" x 27" high						
1900	Minimum	EA	0.889	390	79.00		470
1902	Average	"	1.143	460	100		560
1904	Maximum	"	1.600	540	140		680
1920	60" x 36" high						
1940	Minimum	EA	1.000	400	89.00		490
1942	Average	"	1.333	720	120		840
1944	Maximum	"	1.600	810	140		950
8000	Window frame, milled						
8010	Minimum	LF	0.160	5.73	14.25		20.00
8020	Average	"	0.200	6.39	17.75		24.25
8030	Maximum	"	0.267	9.61	23.75		33.25

SKYLIGHTS

ID Code	Description		Output		Unit Costs			
08 - 62001	**PLASTIC SKYLIGHTS**						**08 - 62001**	
1020	Single thickness, not including mounting curb						
1040	2' x 4'	EA	1.000	430	85.00		520
1050	4' x 4'	"	1.333	580	110		690
1060	5' x 5'	"	2.000	770	170		940
1070	6' x 8'	"	2.667	1,630	230		1,860
1200	Double thickness, not including mounting curb						
1220	2' x 4'	EA	1.000	560	85.00		650
1240	4' x 4'	"	1.333	710	110		820
1260	5' x 5'	"	2.000	1,040	170		1,210
1270	6' x 8'	"	2.667	1,820	230		2,050

HARDWARE

ID Code	Component Descriptions	Unit of Meas.	Manhr / Unit	Material Cost	Labor Cost	Equipment Cost	Total Cost
	Description	**Output**		**Unit Costs**			

08 - 71001 — HINGES — 08 - 71001

ID Code	Component Descriptions	Unit of Meas.	Manhr / Unit	Material Cost	Labor Cost	Equipment Cost	Total Cost
1200	Hinges, material only						
1250	3 x 3 butts, steel, interior, plain bearing	PAIR					21.00
1260	4 x 4 butts, steel, standard	"					31.00
1270	5 x 4-1/2 butts, bronze/s. steel, heavy duty	"					80.00

08 - 71002 — LOCKSETS — 08 - 71002

ID Code	Component Descriptions	Unit of Meas.	Manhr / Unit	Material Cost	Labor Cost	Equipment Cost	Total Cost
1280	Latchset, heavy duty						
1300	Cylindrical	EA	0.500	190	44.50		230
1320	Mortise	"	0.800	200	71.00		270
1325	Lockset, heavy duty						
1330	Cylindrical	EA	0.500	310	44.50		350
1350	Mortise	"	0.800	350	71.00		420
2285	Lockset						
2290	Privacy (bath or bedroom)	EA	0.667	250	59.00		310
2300	Entry lock	"	0.667	280	59.00		340

08 - 71003 — CLOSERS — 08 - 71003

ID Code	Component Descriptions	Unit of Meas.	Manhr / Unit	Material Cost	Labor Cost	Equipment Cost	Total Cost
2600	Door closers						
2610	Standard	EA	1.000	250	89.00		340
2620	Heavy duty	"	1.000	290	89.00		380

08 - 71004 — DOOR TRIM — 08 - 71004

ID Code	Component Descriptions	Unit of Meas.	Manhr / Unit	Material Cost	Labor Cost	Equipment Cost	Total Cost
1600	Panic device						
1610	Mortise	EA	2.000	830	180		1,010
1620	Vertical rod	"	2.000	1,250	180		1,430
1630	Labeled, rim type	"	2.000	860	180		1,040
1640	Mortise	"	2.000	1,130	180		1,310
1650	Vertical rod	"	2.000	1,200	180		1,380

08 - 71006 — WEATHERSTRIPPING — 08 - 71006

ID Code	Component Descriptions	Unit of Meas.	Manhr / Unit	Material Cost	Labor Cost	Equipment Cost	Total Cost
0100	Weatherstrip, head and jamb, metal strip, neoprene						
0140	Standard duty	LF	0.044	5.27	3.94		9.21
0160	Heavy duty	"	0.050	5.86	4.44		10.25
3980	Spring type						
4000	Metal doors	EA	2.000	58.00	180		240
4010	Wood doors	"	2.667	58.00	240		300
4020	Sponge type with adhesive backing	"	0.800	55.00	71.00		130
4500	Thresholds						
4510	Bronze	LF	0.200	57.00	17.75		75.00

HARDWARE

ID Code	Component Descriptions	Unit of Meas.	Manhr / Unit	Material Cost	Labor Cost	Equipment Cost	Total Cost
	Description	**Output**		**Unit Costs**			
08 - 71006	**WEATHERSTRIPPING, Cont'd...**						**08 - 71006**
4515	Aluminum						
4520	Plain	LF	0.200	26.25	17.75		44.00
4525	Vinyl insert	"	0.200	26.75	17.75		44.50
4530	Aluminum with grit	"	0.200	25.50	17.75		43.25
4533	Steel						
4535	Plain	LF	0.200	25.25	17.75		43.00
4540	Interlocking	"	0.667	33.75	59.00		93.00

GLAZING

ID Code	Component Descriptions	Unit of Meas.	Manhr / Unit	Material Cost	Labor Cost	Equipment Cost	Total Cost
08 - 81001	**GLASS GLAZING**						**08 - 81001**
0800	Sheet glass, 1/8" thick	SF	0.044	9.04	3.99		13.00
1020	Plate glass, bronze or grey, 1/4" thick	"	0.073	13.25	6.53		19.75
1040	Clear	"	0.073	10.25	6.53		16.75
1060	Polished	"	0.073	12.25	6.53		18.75
1800	Plexiglass						
2000	1/8" thick	SF	0.073	5.82	6.53		12.25
2020	1/4" thick	"	0.044	10.50	3.99		14.50
3000	Float glass, clear						
3010	3/16" thick	SF	0.067	7.04	5.99		13.00
3020	1/4" thick	"	0.073	7.17	6.53		13.75
3040	3/8" thick	"	0.100	14.50	8.99		23.50
3100	Tinted glass, polished plate, twin ground						
3120	3/16" thick	SF	0.067	9.69	5.99		15.75
3130	1/4" thick	"	0.073	9.69	6.53		16.25
3140	3/8" thick	"	0.100	15.50	8.99		24.50
6800	Insulating glass, two lites, clear float glass						
6840	1/2" thick	SF	0.133	14.00	12.00		26.00
6850	5/8" thick	"	0.160	16.25	14.50		30.75
6860	3/4" thick	"	0.200	17.75	18.00		35.75
6870	7/8" thick	"	0.229	18.75	20.50		39.25
6880	1" thick	"	0.267	25.00	24.00		49.00
6885	Glass seal edge						
6890	3/8" thick	SF	0.133	11.75	12.00		23.75
6895	Tinted glass						
6900	1/2" thick	SF	0.133	29.00	12.00		41.00
6910	1" thick	"	0.267	31.00	24.00		55.00
6920	Tempered, clear						

GLAZING

ID Code	Component Descriptions	Unit of Meas.	Manhr / Unit	Material Cost	Labor Cost	Equipment Cost	Total Cost
		Output		Unit Costs			

ID Code	Description — Component Descriptions	Unit of Meas.	Manhr / Unit	Material Cost	Labor Cost	Equipment Cost	Total Cost
08 - 81001	**GLASS GLAZING, Cont'd...**						**08 - 81001**
6930	1" thick	SF	0.267	47.25	24.00		71.00
7100	Plate mirror glass						
7200	1/4" thick						
7210	15 sf	SF	0.080	15.25	7.19		22.50
7220	Over 15 sf	"	0.073	14.00	6.53		20.50

LOUVERS AND VENTS

ID Code	Component Descriptions	Unit of Meas.	Manhr / Unit	Material Cost	Labor Cost	Equipment Cost	Total Cost
08 - 91001	**VENTS AND WALL LOUVERS**						**08 - 91001**
0100	Block vent, 8"x16"x4" alum., w/screen, mill finish	EA	0.267	210	24.00		230
1200	Standard	"	0.250	120	22.50		140
1210	Vents w/screen, 4" deep, 8" wide, 5" high						
1220	Modular	EA	0.250	140	22.50		160
2000	Aluminum gable louvers	SF	0.133	25.00	12.00		37.00
2020	Vent screen aluminum, 4" wide, continuous	LF	0.027	7.27	2.39		9.66
2260	Aluminum louvers						
4000	Residential use, fixed type, with screen						
4050	8" x 8"	EA	0.400	32.00	36.00		68.00
4060	12" x 12"	"	0.400	35.25	36.00		71.00
4080	12" x 18"	"	0.400	42.25	36.00		78.00
4100	14" x 24"	"	0.400	61.00	36.00		97.00
4120	18" x 24"	"	0.400	68.00	36.00		100
4140	30" x 24"	"	0.444	93.00	40.00		130

DIVISION 09
FINISHES

SUPPORT SYSTEMS

ID Code	Description / Component Descriptions	Output / Unit of Meas.	Output / Manhr / Unit	Unit Costs / Material Cost	Unit Costs / Labor Cost	Unit Costs / Equipment Cost	Unit Costs / Total Cost
09 - 21161	**METAL STUDS**						**09 - 21161**
0060	Studs, non load bearing, galvanized						
0061	1-5/8", 20 ga.						
0062	12" o.c.	SF	0.017	0.91	1.48		2.39
0063	16" o.c.	"	0.013	0.74	1.18		1.92
0064	25 ga.						
0065	12" o.c.	SF	0.017	0.83	1.48		2.31
0066	16" o.c.	"	0.013	0.68	1.18		1.86
0080	2-1/2", 20 ga.						
0100	12" o.c.	SF	0.017	1.02	1.48		2.50
0102	16" o.c.	"	0.013	0.77	1.18		1.95
0110	25 ga.						
0120	12" o.c.	SF	0.017	0.93	1.48		2.41
0122	16" o.c.	"	0.013	0.70	1.18		1.88
0124	24" o.c.	"	0.011	0.46	0.98		1.44
0130	3-5/8", 20 ga.						
0140	12" o.c.	SF	0.020	1.89	1.77		3.66
0142	16" o.c.	"	0.016	1.43	1.42		2.85
0144	24" o.c.	"	0.013	0.94	1.18		2.12
0170	25 ga.						
0180	12" o.c.	SF	0.020	0.98	1.77		2.75
0182	16" o.c.	"	0.016	0.74	1.42		2.16
0184	24" o.c.	"	0.013	0.49	1.18		1.67
0188	4", 20 ga.						
0190	12" o.c.	SF	0.020	1.57	1.77		3.34
0192	16" o.c.	"	0.016	1.18	1.42		2.60
0194	24" o.c.	"	0.013	0.78	1.18		1.96
0198	25 ga.						
0200	12" o.c.	SF	0.020	1.21	1.77		2.98
0202	16" o.c.	"	0.016	0.91	1.42		2.33
0204	24" o.c.	"	0.013	0.60	1.18		1.78
0210	6", 20 ga.						
0220	12" o.c.	SF	0.025	1.69	2.22		3.91
0222	16" o.c.	"	0.020	1.26	1.77		3.03
0224	24" o.c.	"	0.017	0.84	1.48		2.32
0230	25 ga.						
0240	12" o.c.	SF	0.025	1.30	2.22		3.52
0242	16" o.c.	"	0.020	0.98	1.77		2.75
0244	24" o.c.	"	0.017	0.65	1.48		2.13

SUPPORT SYSTEMS

ID Code	Description		Output		Unit Costs			
	Component Descriptions		Unit of Meas.	Manhr / Unit	Material Cost	Labor Cost	Equipment Cost	Total Cost
09 - 21161	**METAL STUDS, Cont'd...**							**09 - 21161**
0300	Tracks, 20 ga.							
0310	1-5/8"		LF	0.012	0.74	1.09		1.83
0320	2-1/2"		"	0.012	0.77	1.09		1.86
0330	3-5/8"		"	0.012	0.89	1.09		1.98
0340	4"		"	0.012	1.20	1.09		2.29
0350	6"		"	0.013	1.26	1.12		2.38
0400	25 ga.							
0410	1-5/8"		LF	0.012	0.68	1.09		1.77
0415	2-1/2"		"	0.012	0.70	1.09		1.79
0420	3-5/8"		"	0.012	0.74	1.09		1.83
0430	4"		"	0.012	0.91	1.09		2.00
0440	6"		"	0.013	0.98	1.12		2.10
0980	Load bearing studs, galvanized							
0990	3-5/8", 16 ga.							
1000	12" o.c.		SF	0.040	2.05	3.55		5.60
1020	16" o.c.		"	0.036	1.63	3.15		4.78
1110	18 ga.							
1130	12" o.c.		SF	0.040	1.54	3.55		5.09
1140	16" o.c.		"	0.036	1.18	3.15		4.33
1145	4", 16 ga.							
1150	12" o.c.		SF	0.040	2.19	3.55		5.74
1160	16" o.c.		"	0.036	1.70	3.15		4.85
1980	6", 16 ga.							
2000	12" o.c.		SF	0.040	2.76	3.55		6.31
2001	16" o.c.		"	0.036	2.13	3.15		5.28
3000	Furring							
3160	On beams and columns							
3170	7/8" channel		LF	0.053	0.65	4.73		5.38
3180	1-1/2" channel		"	0.062	0.77	5.46		6.23
4460	On ceilings							
4470	3/4" furring channels							
4480	12" o.c.		SF	0.033	0.45	2.96		3.41
4490	16" o.c.		"	0.032	0.34	2.84		3.18
4495	24" o.c.		"	0.029	0.22	2.53		2.75
4500	1-1/2" furring channels							
4520	12" o.c.		SF	0.036	0.77	3.22		3.99
4540	16" o.c.		"	0.033	0.58	2.96		3.54
4560	24" o.c.		"	0.031	0.38	2.73		3.11

SUPPORT SYSTEMS

ID Code	Description Component Descriptions	Output		Unit Costs			
		Unit of Meas.	Manhr / Unit	Material Cost	Labor Cost	Equipment Cost	Total Cost
09 - 21161	**METAL STUDS, Cont'd...**						**09 - 21161**
5000	On walls						
5020	3/4" furring channels						
5050	12" o.c.	SF	0.027	0.45	2.36		2.81
5100	16" o.c.	"	0.025	0.34	2.22		2.56
5150	24" o.c.	"	0.024	0.22	2.08		2.30
5200	1-1/2" furring channels						
5210	12" o.c.	SF	0.029	0.77	2.53		3.30
5220	16" o.c.	"	0.027	0.58	2.36		2.94
5230	24" o.c.	"	0.025	0.38	2.22		2.60

LATH AND PLASTER

ID Code	Component Descriptions	Unit of Meas.	Manhr / Unit	Material Cost	Labor Cost	Equipment Cost	Total Cost
09 - 22361	**GYPSUM LATH**						**09 - 22361**
1070	Gypsum lath, 1/2" thick						
1090	Clipped	SY	0.044	5.12	3.94		9.06
1110	Nailed	"	0.050	5.12	4.44		9.56
09 - 22362	**METAL LATH**						**09 - 22362**
0960	Diamond expanded, galvanized						
0980	2.5 lb., on walls						
1010	Nailed	SY	0.100	6.58	8.88		15.50
1030	Wired	"	0.114	6.58	10.25		16.75
1040	On ceilings						
1050	Nailed	SY	0.114	6.58	10.25		16.75
1070	Wired	"	0.133	6.58	11.75		18.25
1980	3.4 lb., on walls						
2000	Nailed	SY	0.100	8.94	8.88		17.75
2020	Wired	"	0.114	8.94	10.25		19.25
2030	On ceilings						
2040	Nailed	SY	0.114	8.94	10.25		19.25
2060	Wired	"	0.133	8.94	11.75		20.75
2064	Flat rib						
2068	2.75 lb., on walls						
2070	Nailed	SY	0.100	6.22	8.88		15.00
2100	Wired	"	0.114	8.94	10.25		19.25
2110	On ceilings						
2120	Nailed	SY	0.114	8.94	10.25		19.25
2140	Wired	"	0.133	8.94	11.75		20.75
2150	3.4 lb., on walls						

LATH AND PLASTER

ID Code	Description / Component Descriptions	Output Unit of Meas.	Manhr / Unit	Material Cost	Labor Cost	Equipment Cost	Total Cost
09 - 22362	**METAL LATH, Cont'd...**						**09 - 22362**
2160	Nailed	SY	0.100	8.94	8.88		17.75
2180	Wired	"	0.114	7.49	10.25		17.75
2190	On ceilings						
2200	Nailed	SY	0.114	7.49	10.25		17.75
2220	Wired	"	0.133	7.49	11.75		19.25
2230	Stucco lath						
2240	1.8 lb.	SY	0.100	7.73	8.88		16.50
2300	3.6 lb.	"	0.100	8.68	8.88		17.50
2310	Paper backed						
2320	Minimum	SY	0.080	6.02	7.10		13.00
2400	Maximum	"	0.114	9.70	10.25		20.00
09 - 22366	**PLASTER ACCESSORIES**						**09 - 22366**
0120	Expansion joint, 3/4", 26 ga., galv.	LF	0.020	1.69	1.77		3.46
2000	Plaster corner beads, 3/4", galvanized	"	0.023	0.47	2.02		2.49
2020	Casing bead, expanded flange, galvanized	"	0.020	0.64	1.77		2.41
2100	Expanded wing, 1-1/4" wide, galvanized	"	0.020	0.75	1.77		2.52
2500	Joint clips for lath	EA	0.004	0.20	0.35		0.55
2580	Metal base, galvanized, 2-1/2" high	LF	0.027	0.86	2.36		3.22
2600	Stud clips for gypsum lath	EA	0.004	0.20	0.35		0.55
2700	Tie wire galvanized, 18 ga., 25 lb. hank	"					54.00
8000	Sound deadening board, 1/4"	SF	0.013	0.36	1.18		1.54
09 - 23001	**PLASTER**						**09 - 23001**
0980	Gypsum plaster, trowel finish, 2 coats						
1000	Ceilings	SY	0.250	6.70	20.75		27.50
1020	Walls	"	0.235	6.70	19.50		26.25
1030	3 coats						
1040	Ceilings	SY	0.348	9.30	28.75		38.00
1060	Walls	"	0.308	9.30	25.50		34.75
1960	Vermiculite plaster						
1980	2 coats						
2000	Ceilings	SY	0.381	5.59	31.50		37.00
2020	Walls	"	0.348	5.59	28.75		34.25
2030	3 coats						
2040	Ceilings	SY	0.471	8.79	39.00		47.75
2060	Walls	"	0.421	8.79	34.75		43.50
5960	Keenes cement plaster						
5980	2 coats						

LATH AND PLASTER

ID Code	Description — Component Descriptions	Output — Unit of Meas.	Output — Manhr / Unit	Unit Costs — Material Cost	Unit Costs — Labor Cost	Unit Costs — Equipment Cost	Unit Costs — Total Cost
09 - 23001	**PLASTER, Cont'd...**						**09 - 23001**
6000	Ceilings	SY	0.308	3.34	25.50		28.75
6020	Walls	"	0.267	3.34	22.00		25.25
6030	3 coats						
6040	Ceilings	SY	0.348	3.12	28.75		31.75
6060	Walls	"	0.308	3.12	25.50		28.50
7000	On columns, add to installation, 50%						
7020	Chases, fascia, and soffits, add to installation, 50%						
7040	Beams, add to installation, 50%						
9000	Patch holes, average size holes						
9020	1 sf to 5 sf						
9022	Minimum	SF	0.133	2.57	11.00		13.50
9024	Average	"	0.160	2.57	13.25		15.75
9026	Maximum	"	0.200	2.57	16.50		19.00
9040	Over 5 sf						
9042	Minimum	SF	0.080	2.57	6.62		9.19
9044	Average	"	0.114	2.57	9.46		12.00
9046	Maximum	"	0.133	2.57	11.00		13.50
9060	Patch cracks						
9062	Minimum	SF	0.027	2.57	2.20		4.77
9064	Average	"	0.040	2.57	3.31		5.88
9066	Maximum	"	0.080	2.57	6.62		9.19
09 - 24001	**PORTLAND CEMENT PLASTER**						**09 - 24001**
2980	Stucco, portland, gray, 3 coat, 1" thick						
3000	Sand finish	SY	0.348	12.25	28.75		41.00
3020	Trowel finish	"	0.364	12.25	30.00		42.25
3030	White cement						
3040	Sand finish	SY	0.364	14.00	30.00		44.00
3060	Trowel finish	"	0.400	14.00	33.00		47.00
3980	Scratch coat						
4000	For ceramic tile	SY	0.080	4.45	6.62		11.00
4020	For quarry tile	"	0.080	4.45	6.62		11.00
5000	Portland cement plaster						
5020	2 coats, 1/2"	SY	0.160	8.84	13.25		22.00
5040	3 coats, 7/8"	"	0.200	10.50	16.50		27.00

GYPSUM BOARD

ID Code	Description — Component Descriptions	Output — Unit of Meas.	Output — Manhr / Unit	Unit Costs — Material Cost	Unit Costs — Labor Cost	Unit Costs — Equipment Cost	Unit Costs — Total Cost
09 - 29001	**GYPSUM BOARD**						**09 - 29001**
1122	Vinyl faced, fastened to metal studs						
1124	1/2" thick	SF	0.010	1.24	0.88		2.12
1126	5/8" thick	"	0.010	1.18	0.88		2.06
1210	Standard drywall, 3/8" thick						
1214	Nailed or screwed to						
1216	Wood or metal framed ceiling	SF	0.008	0.46	0.71		1.17
1218	Columns and beams	"	0.018	0.46	1.57		2.03
1220	Walls	"	0.007	0.46	0.64		1.10
1230	1/2" thick						
1240	Nailed or screwed to						
1250	Wood or metal framed ceiling	SF	0.008	0.43	0.71		1.14
1260	Columns and beams	"	0.018	0.43	1.57		2.00
1270	Walls	"	0.007	0.43	0.64		1.07
1280	5/8" thick						
1290	Nailed or screwed to						
1300	Wood or metal framed ceiling	SF	0.010	0.48	0.88		1.36
1305	Columns and beams	"	0.022	0.48	1.97		2.45
1310	Walls	"	0.009	0.48	0.78		1.26
1320	Lightweight drywall, 1/2" thick						
1380	Nailed or screwed to						
1390	Wood or metal framed ceiling	SF	0.008	0.41	0.71		1.12
1400	Columns and beams	"	0.018	0.41	1.57		1.98
1420	Walls	"	0.007	0.41	0.64		1.05
1430	5/8" thick						
1470	Nailed or screwed to						
1480	Wood or metal framed ceiling	SF	0.010	0.45	0.88		1.33
1490	Columns and beams	"	0.022	0.45	1.97		2.42
1500	Walls	"	0.009	0.45	0.78		1.23
1520	Mold resistant drywall, 1/2" thick						
1580	Nailed or screwed to						
1590	Wood or metal framed ceiling	SF	0.008	0.60	0.71		1.31
1600	Columns and beams	"	0.018	0.60	1.57		2.17
1620	Walls	"	0.007	0.60	0.64		1.24
1700	5/8" thick						
1780	Nailed or screwed to						
1800	Wood or metal framed ceiling	SF	0.010	0.65	0.88		1.53
1820	Columns and beams	"	0.022	0.65	1.97		2.62
1840	Walls	"	0.009	0.65	0.78		1.43

GYPSUM BOARD

	Description		Output		Unit Costs			
ID Code	Component Descriptions		Unit of Meas.	Manhr / Unit	Material Cost	Labor Cost	Equipment Cost	Total Cost

09 - 29001 — GYPSUM BOARD, Cont'd... — 09 - 29001

ID Code	Component Descriptions	Unit of Meas.	Manhr / Unit	Material Cost	Labor Cost	Equipment Cost	Total Cost
2000	Blue Board drywall, 1/2" thick						
2080	Nailed or screwed to						
2100	Wood or metal framed ceiling	SF	0.008	0.52	0.71		1.23
2120	Columns and beams	"	0.018	0.52	1.57		2.09
2140	Walls	"	0.007	0.52	0.64		1.16
2160	5/8" thick						
2240	Nailed or screwed to						
2260	Wood or metal framed ceiling	SF	0.010	0.57	0.88		1.45
2280	Columns and beams	"	0.022	0.57	1.97		2.54
2300	Walls	"	0.009	0.57	0.78		1.35
2500	Soundproof drywall, 1/2" thick						
2580	Nailed or screwed to						
2600	Wood or metal framed ceiling	SF	0.008	1.80	0.71		2.51
2620	Columns and beams	"	0.018	1.80	1.57		3.37
2640	Walls	"	0.007	1.80	0.64		2.44
2660	5/8" thick						
2740	Nailed or screwed to						
2760	Wood or metal framed ceiling	SF	0.010	2.00	0.88		2.88
2780	Columns and beams	"	0.022	2.00	1.97		3.97
2800	Walls	"	0.009	2.00	0.78		2.78
2900	Fire-resistant drywall, 1/2" thick						
2980	Nailed or screwed to						
3000	Wood or metal framed ceiling	SF	0.008	0.52	0.71		1.23
3020	Columns and beams	"	0.018	0.52	1.57		2.09
3040	Walls	"	0.007	0.52	0.64		1.16
3060	5/8" thick						
3140	Nailed or screwed to						
3160	Wood or metal framed ceiling	SF	0.010	0.57	0.88		1.45
3180	Columns and beams	"	0.022	0.57	1.97		2.54
3200	Walls	"	0.009	0.57	0.78		1.35
3220	Impact-resistant drywall, 5/8" thick						
3300	Nailed or screwed to						
3330	Wood or metal framed ceiling	SF	0.010	0.94	0.88		1.82
3340	Columns and beams	"	0.022	0.94	1.97		2.91
3360	Walls	"	0.009	0.94	0.78		1.72
3370	High-flex drywall, 1/4" thick, nailed or screwed						
3380	Walls	SF	0.009	0.48	0.78		1.26
3400	Shaftwall drywall, 1" thick, nailed or screwed						

GYPSUM BOARD

ID Code	Description		Output		Unit Costs			
	Component Descriptions	Unit of Meas.	Manhr / Unit	Material Cost	Labor Cost	Equipment Cost	Total Cost
09 - 29001	**GYPSUM BOARD, Cont'd...**						**09 - 29001**
3420	Walls	SF	0.011	0.89	1.01		1.90
3500	Stretch drywall, 1/2" thick						
3530	Nailed or screwed to						
3540	Wood or metal framed ceiling	SF	0.009	0.34	0.78		1.12
3550	Walls	"	0.008	0.34	0.71		1.05
4000	Tile backer board, screwed to wall						
4020	1/4" thick	SF	0.032	1.02	2.84		3.86
4040	7/16" thick	"	0.032	1.14	2.84		3.98
4060	1/2" thick	"	0.032	0.97	2.84		3.81
4080	5/8" thick	"	0.032	1.13	2.84		3.97
5020	Casing bead						
5022	Minimum	LF	0.023	0.17	2.02		2.19
5024	Average	"	0.027	0.18	2.36		2.54
5026	Maximum	"	0.040	0.23	3.55		3.78
5040	Corner bead						
5042	Minimum	LF	0.023	0.18	2.02		2.20
5044	Average	"	0.027	0.23	2.36		2.59
5046	Maximum	"	0.040	0.28	3.55		3.83
6000	Taping and finishing joints						
6020	Level 1	SF	0.005	0.05	0.47		0.52
6040	Level 2	"	0.006	0.05	0.54		0.59
6060	Level 3	"	0.007	0.07	0.59		0.66
6080	Level 4	"	0.008	0.06	0.71		0.77
6100	Level 5 (includes skim coat)	"	0.008	0.11	0.71		0.82
6120	Add to above for Special Textures						
6200	Orange Peel	SF	0.013	0.40	1.18		1.58
6220	Santa-Fe	"	0.016	0.57	1.42		1.99
6240	Skip Trowel	"	0.015	0.51	1.29		1.80
6260	Smooth	"	0.013	0.40	1.18		1.58
6280	Splatter Knockdown	"	0.015	0.51	1.29		1.80
6290	Swirl	"	0.018	0.51	1.57		2.08
6300	Rosebud	"	0.018	0.51	1.57		2.08
6320	Popcorn	"	0.015	0.51	1.29		1.80

TILE

ID Code	Component Descriptions	Unit of Meas.	Manhr / Unit	Material Cost	Labor Cost	Equipment Cost	Total Cost
09 - 30131	**CERAMIC TILE**						**09 - 30131**
0980	Glazed wall tile, 4-1/4" x 4-1/4"						
1000	Minimum	SF	0.057	2.57	4.84		7.41
1020	Average	"	0.067	4.08	5.65		9.73
1040	Maximum	"	0.080	14.50	6.78		21.25
2960	Base, 4-1/4" high						
2980	Minimum	LF	0.100	4.72	8.47		13.25
3000	Average	"	0.100	5.48	8.47		14.00
3040	Maximum	"	0.100	7.25	8.47		15.75
6100	Unglazed floor tile						
6120	Portland cem., cushion edge, face mtd						
6140	1" x 1"	SF	0.073	17.75	6.16		24.00
6150	2" x 2"	"	0.067	18.75	5.65		24.50
6162	4" x 4"	"	0.067	17.50	5.65		23.25
6164	6" x 6"	"	0.057	6.26	4.84		11.00
6166	12" x 12"	"	0.050	5.51	4.23		9.74
6168	16" x 16"	"	0.044	4.78	3.76		8.54
6170	18" x 18"	"	0.040	4.63	3.39		8.02
6200	Adhesive bed, with white grout						
6220	1" x 1"	SF	0.073	14.75	6.16		21.00
6230	2" x 2"	"	0.067	15.75	5.65		21.50
6260	4" x 4"	"	0.067	15.75	5.65		21.50
6262	6" x 6"	"	0.057	5.22	4.84		10.00
6264	12" x 12"	"	0.050	4.58	4.23		8.81
6266	16" x 16"	"	0.044	3.96	3.76		7.72
6268	18" x 18"	"	0.040	3.85	3.39		7.24
6300	Organic adhesive bed, thin set, back mounted						
6320	1" x 1"	SF	0.073	14.75	6.16		21.00
6350	2" x 2"	"	0.067	17.25	5.65		23.00
6360	For group 2 colors, add to material, 10%						
6370	For group 3 colors, add to material, 20%						
6380	For abrasive surface, add to material, 25%						
6382	Porcelain floor tile						
6384	1" x 1"	SF	0.073	10.50	6.16		16.75
6386	2" x 2"	"	0.070	9.55	5.89		15.50
6388	4" x 4"	"	0.067	8.87	5.65		14.50
6390	6" x 6"	"	0.057	3.19	4.84		8.03
6392	12" x 12"	"	0.050	2.87	4.23		7.10
6394	16" x 16"	"	0.044	2.28	3.76		6.04

TILE

ID Code	Description Component Descriptions	Output		Unit Costs			
		Unit of Meas.	Manhr / Unit	Material Cost	Labor Cost	Equipment Cost	Total Cost
09 - 30131	**CERAMIC TILE, Cont'd...**						**09 - 30131**
6396	18" x 18"	SF	0.040	2.15	3.39		5.54
6400	Unglazed wall tile						
6420	Organic adhesive, face mounted cushion edge						
6425	1" x 1"						
6430	Minimum	SF	0.067	6.87	5.65		12.50
6432	Average	"	0.073	9.00	6.16		15.25
6434	Maximum	"	0.080	13.50	6.78		20.25
6448	2" x 2"						
6450	Minimum	SF	0.062	7.94	5.21		13.25
6452	Average	"	0.067	9.00	5.65		14.75
6454	Maximum	"	0.073	14.75	6.16		21.00
6500	Back mounted						
6510	1" x 1"						
6520	Minimum	SF	0.067	6.87	5.65		12.50
6522	Average	"	0.073	9.00	6.16		15.25
6524	Maximum	"	0.080	13.50	6.78		20.25
6538	2" x 2"						
6540	Minimum	SF	0.062	7.94	5.21		13.25
6542	Average	"	0.067	9.00	5.65		14.75
6544	Maximum	"	0.073	14.75	6.16		21.00
6600	For glazed finish, add to material, 25%						
6620	For glazed mosaic, add to material, 100%						
6630	For metallic colors, add to material, 125%						
6640	For exterior wall use, add to total, 25%						
6650	For exterior soffit, add to total, 25%						
6660	For portland cement bed, add to total, 25%						
6670	For dry set portland cement bed, add to total, 10%						
8990	Ceramic accessories						
9000	Towel bar, 24" long						
9002	Minimum	EA	0.320	19.00	27.00		46.00
9004	Average	"	0.400	23.25	34.00		57.00
9006	Maximum	"	0.533	63.00	45.25		110
9020	Soap dish						
9022	Minimum	EA	0.533	8.93	45.25		54.00
9024	Average	"	0.667	12.00	57.00		69.00
9026	Maximum	"	0.800	32.00	68.00		100

TILE

	Description	Output		Unit Costs			
ID Code	Component Descriptions	Unit of Meas.	Manhr / Unit	Material Cost	Labor Cost	Equipment Cost	Total Cost

09 - 30161　　QUARRY TILE　　09 - 30161

1060	Floor						
1080	4 x 4 x 1/2"	SF	0.107	6.93	9.04		16.00
1100	6 x 6 x 1/2"	"	0.100	6.79	8.47		15.25
1120	6 x 6 x 3/4"	"	0.100	8.43	8.47		17.00
1122	12 x 12 x 3/4"	"	0.089	11.75	7.53		19.25
1124	16 x 1 6 x 3/4"	"	0.080	8.27	6.78		15.00
1126	18 x 18 x 3/4"	"	0.067	5.82	5.65		11.50
1150	Medallion						
1160	36" dia.	EA	2.000	610	170		780
1162	48" dia.	"	2.000	720	170		890
1200	Wall, applied to 3/4" portland cement bed						
1220	4 x 4 x 1/2"	SF	0.160	6.16	13.50		19.75
1240	6 x 6 x 3/4"	"	0.133	6.89	11.25		18.25
1320	Cove base						
1330	5 x 6 x 1/2" straight top	LF	0.133	7.03	11.25		18.25
1340	6 x 6 x 3/4" round top	"	0.133	6.52	11.25		17.75
1345	Moldings						
1350	2 x 12	LF	0.080	11.25	6.78		18.00
1352	4 x 12	"	0.080	17.50	6.78		24.25
1360	Stair treads 6 x 6 x 3/4"	"	0.200	9.63	17.00		26.75
1380	Window sill 6 x 8 x 3/4"	"	0.160	8.79	13.50		22.25
1400	For abrasive surface, add to material, 25%						

ACOUSTICAL TREATMENT

09 - 51001　　CEILINGS AND WALLS　　09 - 51001

1400	Acoustical panels, suspension system not included						
1420	Fiberglass panels						
1500	5/8" thick						
1560	2' x 2'	SF	0.011	1.96	1.01		2.97
1580	2' x 4'	"	0.009	1.62	0.78		2.40
1590	3/4" thick						
1600	2' x 2'	SF	0.011	2.60	1.01		3.61
1620	2' x 4'	"	0.009	2.52	0.78		3.30
1630	Glass cloth faced fiberglass panels						
1660	3/4" thick	SF	0.013	3.71	1.18		4.89
1680	1" thick	"	0.013	4.13	1.18		5.31
1690	Mineral fiber panels						

ACOUSTICAL TREATMENT

ID Code	Description — Component Descriptions	Output — Unit of Meas.	Output — Manhr / Unit	Unit Costs — Material Cost	Unit Costs — Labor Cost	Unit Costs — Equipment Cost	Unit Costs — Total Cost
09 - 51001	**CEILINGS AND WALLS, Cont'd...**						**09 - 51001**
1700	5/8" thick						
1720	2' x 2'	SF	0.011	1.66	1.01		2.67
1740	2' x 4'	"	0.009	1.66	0.78		2.44
1750	3/4" thick						
1760	2' x 2'	SF	0.011	2.60	1.01		3.61
1780	2' x 4'	"	0.009	2.55	0.78		3.33
1820	Wood fiber panels						
1840	1/2" thick						
1850	2' x 2'	SF	0.011	2.80	1.01		3.81
1860	2' x 4'	"	0.009	2.80	0.78		3.58
1870	5/8" thick						
1880	2' x 2'	SF	0.011	3.22	1.01		4.23
1890	2' x 4'	"	0.009	3.22	0.78		4.00
2090	Acoustical tiles, suspension system not included						
2100	Fiberglass tile, 12" x 12"						
3040	5/8" thick	SF	0.015	2.11	1.29		3.40
3060	3/4" thick	"	0.018	2.45	1.57		4.02
3080	Glass cloth faced fiberglass tile						
3100	3/4" thick	SF	0.018	3.93	1.57		5.50
3120	3" thick	"	0.020	4.40	1.77		6.17
3130	Mineral fiber tile, 12" x 12"						
3140	5/8" thick						
3160	Standard	SF	0.016	1.10	1.42		2.52
3170	Vinyl faced	"	0.016	2.20	1.42		3.62
3180	3/4" thick						
3190	Standard	SF	0.016	1.61	1.42		3.03
3200	Vinyl faced	"	0.016	2.81	1.42		4.23
5420	Ceiling suspension systems						
5440	T-bar system						
5510	2' x 4'	SF	0.008	1.34	0.71		2.05
5520	2' x 2'	"	0.009	1.45	0.78		2.23
5530	Concealed Z-bar suspension system, 12" module	"	0.013	1.38	1.18		2.56
5550	For 1-1/2" carrier channels, 4' o.c., add	"					0.44
5560	Carrier channel for recessed light fixtures	"					0.80
5820	T-bar system grid components						
5822	2' Cross Tee	EA	0.010	2.32	0.88		3.20
5824	4' Cross Tee	"	0.011	4.64	0.94		5.58
5826	Main Beam	LF	0.011	1.27	1.01		2.28

ACOUSTICAL TREATMENT

ID Code	Component Descriptions	Unit of Meas.	Manhr / Unit	Material Cost	Labor Cost	Equipment Cost	Total Cost
09 - 51001	**CEILINGS AND WALLS, Cont'd...**						**09 - 51001**
5828	Wall Molding	LF	0.008	0.69	0.71		1.40
5900	Ceiling tile, quick costs						
5902	Minimum	SF	0.016	1.85	1.42		3.27
5904	Average	"	0.018	2.61	1.57		4.18
5906	Maximum	"	0.020	6.49	1.77		8.26

FLOORING

ID Code	Component Descriptions	Unit of Meas.	Manhr / Unit	Material Cost	Labor Cost	Equipment Cost	Total Cost
09 - 63161	**UNIT MASONRY FLOORING**						**09 - 63161**
1000	Clay brick						
1020	9 x 4-1/2 x 3" thick						
1040	Glazed	SF	0.067	7.44	5.92		13.25
1060	Unglazed	"	0.067	7.13	5.92		13.00
1070	8 x 4 x 3/4" thick						
1080	Glazed	SF	0.070	6.73	6.17		13.00
1100	Unglazed	"	0.070	6.42	6.17		12.50
1140	For herringbone pattern, add to labor, 15%						
09 - 64001	**WOOD FLOORING**						**09 - 64001**
0100	Wood strip flooring, unfinished						
1000	Fir floor						
1010	C and better						
1020	Vertical grain	SF	0.027	3.60	2.36		5.96
1040	Flat grain	"	0.027	4.51	2.36		6.87
1060	Oak floor						
1080	Minimum	SF	0.038	3.82	3.38		7.20
1100	Average	"	0.038	5.25	3.38		8.63
1120	Maximum	"	0.038	7.61	3.38		11.00
1200	Maple floor						
1220	25/32" x 2-1/4"						
1240	Minimum	SF	0.038	5.45	3.38		8.83
1260	Maximum	"	0.038	7.72	3.38		11.00
1280	33/32" x 3-1/4"						
1300	Minimum	SF	0.038	7.60	3.38		11.00
1320	Maximum	"	0.038	8.59	3.38		12.00
1340	Added costs						
1350	For factory finish, add to material, 10%						
1355	For random width floor, add to total, 20%						
1360	For simulated pegs, add to total, 10%						

FLOORING

ID Code	Component Descriptions	Unit of Meas.	Manhr / Unit	Material Cost	Labor Cost	Equipment Cost	Total Cost
	Description	**Output**		**Unit Costs**			
09 - 64001	**WOOD FLOORING, Cont'd...**						**09 - 64001**
1500	Wood block industrial flooring						
1510	Creosoted						
1520	2" thick	SF	0.021	4.30	1.86		6.16
1540	2-1/2" thick	"	0.025	4.47	2.22		6.69
1560	3" thick	"	0.027	4.64	2.36		7.00
2500	Parquet, 5/16", white oak						
2520	Finished	SF	0.040	10.25	3.55		13.75
2540	Unfinished	"	0.040	4.99	3.55		8.54
3000	Gym floor, 2 ply felt, 25/32" maple, finished, in mastic	"	0.044	8.80	3.94		12.75
3020	Over wood sleepers	"	0.050	8.96	4.44		13.50
9020	Finishing, sand, fill, finish, and wax	"	0.020	0.68	1.77		2.45
9100	Refinish sand, seal, and 2 coats of polyurethane	"	0.027	1.19	2.36		3.55
9540	Clean and wax floors	"	0.004	0.24	0.35		0.59
09 - 65131	**RESILIENT BASE AND ACCESSORIES**						**09 - 65131**
1000	Wall base, vinyl						
1130	4" high	LF	0.027	1.70	2.36		4.06
1140	6" high	"	0.027	2.31	2.36		4.67
09 - 65161	**RESILIENT SHEET FLOORING**						**09 - 65161**
0980	Vinyl sheet flooring						
1000	Minimum	SF	0.008	4.34	0.71		5.05
1002	Average	"	0.010	9.27	0.84		10.00
1004	Maximum	"	0.013	17.75	1.18		19.00
1020	Cove, to 6"	LF	0.016	5.08	1.42		6.50
2000	Fluid applied resilient flooring						
2020	Polyurethane, poured in place, 3/8" thick	SF	0.067	11.00	5.92		17.00
6200	Vinyl sheet goods, backed						
6220	0.070" thick	SF	0.010	4.22	0.88		5.10
6240	0.093" thick	"	0.010	6.54	0.88		7.42
6260	0.125" thick	"	0.010	7.55	0.88		8.43
6280	0.250" thick	"	0.010	8.68	0.88		9.56
09 - 65191	**RESILIENT TILE FLOORING**						**09 - 65191**
1020	Solid vinyl tile, 1/8" thick, 12" x 12"						
1040	Marble patterns	SF	0.020	5.28	1.77		7.05
1060	Solid colors	"	0.020	6.85	1.77		8.62
1080	Travertine patterns	"	0.020	7.70	1.77		9.47
2000	Conductive resilient flooring, vinyl tile						

FLOORING

ID Code	Component Descriptions	Unit of Meas.	Manhr / Unit	Material Cost	Labor Cost	Equipment Cost	Total Cost
	Description	**Output**		**Unit Costs**			
09 - 65191	**RESILIENT TILE FLOORING, Cont'd...**						**09 - 65191**
2040	1/8" thick, 12" x 12"	SF	0.023	9.18	2.02		11.25
09 - 66131	**TERRAZZO**						**09 - 66131**
1100	Floors on concrete, 1-3/4" thick, 5/8" topping						
1120	Gray cement	SF	0.114	7.18	9.46		16.75
1140	White cement	"	0.114	7.50	9.46		17.00
1200	Sand cushion, 3" thick, 5/8" top, 1/4"						
1220	Gray cement	SF	0.133	8.48	11.00		19.50
1240	White cement	"	0.133	8.81	11.00		19.75
1260	Monolithic terrazzo, 3-1/2" base slab, 5/8" topping	"	0.100	6.01	8.28		14.25
1280	Terrazzo wainscot, cast-in-place, 1/2" thick	"	0.200	7.31	16.50		23.75
1300	Base, cast-in-place, terrazzo cove type, 6" high	LF	0.114	9.19	9.46		18.75
1320	Curb, cast-in-place, 6" wide x 6" high, polished top	"	0.400	8.35	33.00		41.25
1340	For venetian type terrazzo, add to material, 10%						
1360	For abrasive heavy duty terrazzo, add to material,						
1400	Divider strips						
1500	Zinc	LF					1.50
1510	Brass	"					2.80
1560	Stairs, cast-in-place, topping on concrete or metal						
1620	1-1/2" thick treads, 12" wide	LF	0.400	5.91	33.00		39.00
1640	Combined tread and riser	"	1.000	8.88	83.00		92.00
1680	Precast terrazzo, thin set						
1690	Terrazzo tiles, non-slip surface						
2120	9" x 9" x 1" thick	SF	0.114	17.00	9.46		26.50
2130	12" x 12"						
2140	1" thick	SF	0.107	18.25	8.83		27.00
2160	1-1/2" thick	"	0.114	19.00	9.46		28.50
2180	18" x 18" x 1-1/2" thick	"	0.114	25.00	9.46		34.50
2200	24" x 24" x 1-1/2" thick	"	0.094	32.00	7.79		39.75
2400	For white cement, add to material, 10%						
2800	For venetian type terrazzo, add to material, 25%						
3000	Terrazzo wainscot						
3020	12" x 12" x 1" thick	SF	0.200	9.31	16.50		25.75
3040	18" x 18" x 1-1/2" thick	"	0.229	15.25	19.00		34.25
3060	Base						
3080	6" high						
3220	Straight	LF	0.062	13.25	5.09		18.25
3240	Coved	"	0.062	15.75	5.09		20.75

FLOORING

		Output		Unit Costs			
ID Code	Component Descriptions	Unit of Meas.	Manhr / Unit	Material Cost	Labor Cost	Equipment Cost	Total Cost
09 - 66131	**TERRAZZO, Cont'd...**						**09 - 66131**
3260	8" high						
3280	Straight	LF	0.067	15.00	5.52		20.50
3300	Coved	"	0.067	17.50	5.52		23.00
3310	Terrazzo curbs						
3320	8" wide x 8" high	LF	0.320	34.75	26.50		61.00
3340	6" wide x 6" high	"	0.267	31.50	22.00		54.00
3400	Precast terrazzo stair treads, 12" wide						
3410	1-1/2" thick						
3420	Diamond pattern	LF	0.145	42.00	12.00		54.00
3430	Non-slip surface	"	0.145	44.00	12.00		56.00
3440	2" thick						
3450	Diamond pattern	LF	0.145	44.00	12.00		56.00
3460	Non-slip surface	"	0.160	46.25	13.25		60.00
3480	Stair risers, 1" thick to 6" high						
3520	Straight sections	LF	0.080	14.00	6.62		20.50
3530	Cove sections	"	0.080	16.50	6.62		23.00
3600	Combined tread and riser						
3620	Straight sections						
3640	1-1/2" tread, 3/4" riser	LF	0.229	61.00	19.00		80.00
3660	3" tread, 1" riser	"	0.229	72.00	19.00		91.00
3680	Curved sections						
3700	2" tread, 1" riser	LF	0.267	77.00	22.00		99.00
3720	3" tread, 1" riser	"	0.267	80.00	22.00		100
3800	Stair stringers, notched for treads and risers						
3820	1" thick	LF	0.200	36.50	16.50		53.00
3840	2" thick	"	0.267	38.00	22.00		60.00
3860	Landings, structural, nonslip						
3870	1-1/2" thick	SF	0.133	34.50	11.00		45.50
3880	3" thick	"	0.160	48.25	13.25		62.00

CARPET

	Description	Output		Unit Costs			
ID Code	Component Descriptions	Unit of Meas.	Manhr / Unit	Material Cost	Labor Cost	Equipment Cost	Total Cost
09 - 68001		**CARPET PADDING**					**09 - 68001**
1000	Carpet padding						
1005	Foam rubber, waffle type, 0.3" thick	SY	0.040	6.91	3.55		10.50
1010	Jute padding						
1020	Minimum	SY	0.036	5.86	3.22		9.08
1022	Average	"	0.040	7.63	3.55		11.25
1024	Maximum	"	0.044	11.50	3.94		15.50
1030	Sponge rubber cushion						
1040	Minimum	SY	0.036	5.56	3.22		8.78
1042	Average	"	0.040	7.40	3.55		11.00
1044	Maximum	"	0.044	10.50	3.94		14.50
1050	Urethane cushion, 3/8" thick						
1060	Minimum	SY	0.036	5.56	3.22		8.78
1062	Average	"	0.040	6.48	3.55		10.00
1064	Maximum	"	0.044	8.45	3.94		12.50
09 - 68002		**CARPET**					**09 - 68002**
0990	Carpet, acrylic						
1000	24 oz., light traffic	SY	0.089	19.00	7.89		27.00
1020	28 oz., medium traffic	"	0.089	22.75	7.89		30.75
2010	Nylon						
2020	15 oz., light traffic	SY	0.089	26.50	7.89		34.50
2040	28 oz., medium traffic	"	0.089	34.25	7.89		42.25
2110	Nylon						
2120	28 oz., medium traffic	SY	0.089	33.00	7.89		41.00
2140	35 oz., heavy traffic	"	0.089	40.00	7.89		48.00
2145	Wool						
2150	30 oz., medium traffic	SY	0.089	82.00	7.89		90.00
2160	36 oz., medium traffic	"	0.089	86.00	7.89		94.00
2180	42 oz., heavy traffic	"	0.089	110	7.89		120
3000	Carpet tile						
3020	Foam backed						
3022	Minimum	SF	0.016	4.38	1.42		5.80
3024	Average	"	0.018	5.07	1.57		6.64
3026	Maximum	"	0.020	8.04	1.77		9.81
3040	Tufted loop or shag						
3042	Minimum	SF	0.016	4.75	1.42		6.17
3044	Average	"	0.018	5.73	1.57		7.30
3046	Maximum	"	0.020	9.22	1.77		11.00

CARPET

ID Code	Component Descriptions	Unit of Meas.	Manhr / Unit	Material Cost	Labor Cost	Equipment Cost	Total Cost
	Description	**Output**		**Unit Costs**			
09 - 68002	**CARPET, Cont'd...**						**09 - 68002**
8980	Clean and vacuum carpet						
9000	Minimum	SY	0.004	0.39	0.27		0.66
9020	Average	"	0.005	0.60	0.47		1.07
9040	Maximum	"	0.008	0.82	0.71		1.53

WALL COVERING

ID Code	Component Descriptions	Unit of Meas.	Manhr / Unit	Material Cost	Labor Cost	Equipment Cost	Total Cost
09 - 72001	**WALL COVERING**						**09 - 72001**
0900	Vinyl wall covering						
1000	Medium duty	SF	0.011	1.16	0.84		2.00
1010	Heavy duty	"	0.013	2.41	0.98		3.39
1020	Over pipes and irregular shapes						
1030	Lightweight, 13 oz.	SF	0.016	2.02	1.18		3.20
1040	Medium weight, 25 oz.	"	0.018	2.41	1.31		3.72
1060	Heavyweight, 34 oz.	"	0.020	2.95	1.48		4.43
1080	Cork wall covering						
1100	1' x 1' squares						
1140	1/4" thick	SF	0.020	6.03	1.48		7.51
1160	1/2" thick	"	0.020	7.66	1.48		9.14
1180	3/4" thick	"	0.020	8.63	1.48		10.00
1190	Wall fabrics						
1200	Natural fabrics, grass cloths						
1220	Minimum	SF	0.012	1.75	0.91		2.66
1240	Average	"	0.013	1.95	0.98		2.93
1260	Maximum	"	0.016	6.53	1.18		7.71
1280	Flexible gypsum coated wall fabric, fire resistant	"	0.008	1.97	0.59		2.56
2000	Vinyl corner guards						
2020	3/4" x 3/4" x 8'	EA	0.100	9.26	7.41		16.75
2040	2-3/4" x 2-3/4" x 4'	"	0.100	5.46	7.41		12.75

PAINT

ID Code	Component Descriptions	Unit of Meas.	Manhr / Unit	Material Cost	Labor Cost	Equipment Cost	Total Cost
09 - 91001	**PAINTING PREPARATION**						**09 - 91001**
1000	Dropcloths						
1050	Minimum	SF	0.001	0.17	0.03		0.20
1100	Average	"	0.001	0.19	0.04		0.23
1150	Maximum	"	0.001	0.39	0.06		0.45
1200	Masking						

PAINT

ID Code	Component Descriptions	Unit of Meas.	Manhr / Unit	Material Cost	Labor Cost	Equipment Cost	Total Cost
09 - 91001	**PAINTING PREPARATION, Cont'd...**						**09 - 91001**
1250	Paper and tape						
1300	Minimum	LF	0.008	0.04	0.59		0.63
1350	Average	"	0.010	0.07	0.74		0.81
1400	Maximum	"	0.013	0.08	0.98		1.06
1450	Doors						
1500	Minimum	EA	0.100	0.05	7.41		7.46
1550	Average	"	0.133	0.07	9.88		9.95
1600	Maximum	"	0.178	0.08	13.25		13.25
1650	Windows						
1700	Minimum	EA	0.100	0.05	7.41		7.46
1750	Average	"	0.133	0.07	9.88		9.95
1800	Maximum	"	0.178	0.08	13.25		13.25
2000	Sanding						
2050	Walls and flat surfaces						
2100	Minimum	SF	0.005		0.39		0.39
2150	Average	"	0.007		0.49		0.49
2200	Maximum	"	0.008		0.59		0.59
2250	Doors and windows						
2300	Minimum	EA	0.133		9.88		9.88
2350	Average	"	0.200		14.75		14.75
2400	Maximum	"	0.267		19.75		19.75
2450	Trim						
2500	Minimum	LF	0.010		0.74		0.74
2550	Average	"	0.013		0.98		0.98
2600	Maximum	"	0.018		1.31		1.31
2650	Puttying						
2700	Minimum	SF	0.012	0.02	0.91		0.93
2750	Average	"	0.016	0.03	1.18		1.21
2800	Maximum	"	0.020	0.05	1.48		1.53
09 - 91009	**PAINT**						**09 - 91009**
0830	Paint, enamel						
0850	600 sf per gal.	GAL					58.00
0900	550 sf per gal.	"					54.00
1000	500 sf per gal.	"					38.25
1020	450 sf per gal.	"					35.75
1060	350 sf per gal.	"					34.50
1100	Filler, 60 sf per gal.	"					41.00

PAINT

	Description	Output		Unit Costs			
ID Code	Component Descriptions	Unit of Meas.	Manhr / Unit	Material Cost	Labor Cost	Equipment Cost	Total Cost
09 - 91009	**PAINT, Cont'd...**						**09 - 91009**
1160	Latex, 400 sf per gal.	GAL					38.25
1170	Aluminum						
1180	400 sf per gal.	GAL					51.00
1190	500 sf per gal.	"					82.00
1200	Red lead, 350 sf per gal.	"					72.00
1220	Primer						
1240	400 sf per gal.	GAL					34.25
1250	300 sf per gal.	"					34.50
1280	Latex base, interior, white	"					38.25
1480	Sealer and varnish						
1500	400 sf per gal.	GAL					35.75
1520	425 sf per gal.	"					51.00
1540	600 sf per gal.	"					67.00
09 - 91130	**EXT. PAINTING, SITEWORK**						**09 - 91130**
3000	Concrete Block						
3020	Roller						
3040	First Coat						
3060	Minimum	SF	0.004	0.21	0.29		0.50
3080	Average	"	0.005	0.22	0.39		0.61
3100	Maximum	"	0.008	0.23	0.59		0.82
3120	Second Coat						
3140	Minimum	SF	0.003	0.21	0.24		0.45
3160	Average	"	0.004	0.22	0.32		0.54
3180	Maximum	"	0.007	0.23	0.49		0.72
3200	Spray						
3220	First Coat						
3240	Minimum	SF	0.002	0.16	0.16		0.32
3260	Average	"	0.003	0.18	0.19		0.37
3280	Maximum	"	0.003	0.19	0.22		0.41
3300	Second Coat						
3320	Minimum	SF	0.001	0.16	0.10		0.26
3340	Average	"	0.002	0.18	0.13		0.31
3360	Maximum	"	0.003	0.19	0.18		0.37
3500	Fences, Chain Link						
3700	Roller						
3720	First Coat						
3740	Minimum	SF	0.006	0.14	0.42		0.56

PAINT

ID Code	Component Descriptions	Unit of Meas.	Manhr / Unit	Material Cost	Labor Cost	Equipment Cost	Total Cost
	Description	**Output**		**Unit Costs**			

09 - 91130   **EXT. PAINTING, SITEWORK, Cont'd...**   **09 - 91130**

ID Code	Component Descriptions	Unit of Meas.	Manhr / Unit	Material Cost	Labor Cost	Equipment Cost	Total Cost
3760	Average	SF	0.007	0.15	0.49		0.64
3780	Maximum	"	0.008	0.16	0.56		0.72
3800	Second Coat						
3820	Minimum	SF	0.003	0.14	0.24		0.38
3840	Average	"	0.004	0.15	0.29		0.44
3860	Maximum	"	0.005	0.16	0.37		0.53
3880	Spray						
3900	First Coat						
3920	Minimum	SF	0.003	0.11	0.18		0.29
3940	Average	"	0.003	0.12	0.21		0.33
3960	Maximum	"	0.003	0.14	0.24		0.38
3980	Second Coat						
4000	Minimum	SF	0.002	0.11	0.14		0.25
4060	Average	"	0.002	0.12	0.16		0.28
4080	Maximum	"	0.003	0.14	0.18		0.32
4200	Fences, Wood or Masonry						
4220	Brush						
4240	First Coat						
4260	Minimum	SF	0.008	0.21	0.62		0.83
4280	Average	"	0.010	0.22	0.74		0.96
4300	Maximum	"	0.013	0.23	0.98		1.21
4320	Second Coat						
4340	Minimum	SF	0.005	0.21	0.37		0.58
4360	Average	"	0.006	0.22	0.45		0.67
4380	Maximum	"	0.008	0.23	0.59		0.82
4400	Roller						
4420	First Coat						
4440	Minimum	SF	0.004	0.21	0.32		0.53
4460	Average	"	0.005	0.22	0.39		0.61
4480	Maximum	"	0.006	0.23	0.45		0.68
4500	Second Coat						
4520	Minimum	SF	0.003	0.21	0.22		0.43
4540	Average	"	0.004	0.22	0.28		0.50
4560	Maximum	"	0.005	0.23	0.37		0.60
4580	Spray						
4600	First Coat						
4620	Minimum	SF	0.003	0.16	0.21		0.37
4640	Average	"	0.004	0.18	0.26		0.44

PAINT

ID Code	Description		Output		Unit Costs			
	Component Descriptions	Unit of Meas.	Manhr / Unit	Material Cost	Labor Cost	Equipment Cost	Total Cost
09 - 91130	**EXT. PAINTING, SITEWORK, Cont'd...**						**09 - 91130**
4660	Maximum	SF	0.005	0.19	0.37		0.56
4680	Second Coat						
4700	Minimum	SF	0.002	0.16	0.14		0.30
4760	Average	"	0.003	0.18	0.18		0.36
4780	Maximum	"	0.003	0.19	0.24		0.43
09 - 91131	**EXT. PAINTING, BUILDINGS**						**09 - 91131**
1200	Decks, Wood, Stained						
1220	Brush						
1240	First Coat						
1260	Minimum	SF	0.004	0.16	0.29		0.45
1280	Average	"	0.004	0.18	0.32		0.50
1300	Maximum	"	0.005	0.19	0.37		0.56
1320	Second Coat						
1340	Minimum	SF	0.003	0.16	0.21		0.37
1360	Average	"	0.003	0.18	0.22		0.40
1380	Maximum	"	0.003	0.19	0.24		0.43
1400	Roller						
1420	First Coat						
1440	Minimum	SF	0.003	0.16	0.21		0.37
1460	Average	"	0.003	0.18	0.22		0.40
1480	Maximum	"	0.003	0.19	0.24		0.43
1500	Second Coat						
1520	Minimum	SF	0.003	0.16	0.18		0.34
1540	Average	"	0.003	0.18	0.19		0.37
1560	Maximum	"	0.003	0.19	0.22		0.41
1580	Spray						
1600	First Coat						
1620	Minimum	SF	0.003	0.14	0.18		0.32
1640	Average	"	0.003	0.15	0.19		0.34
1660	Maximum	"	0.003	0.16	0.22		0.38
1680	Second Coat						
1700	Minimum	SF	0.002	0.14	0.16		0.30
1720	Average	"	0.002	0.15	0.17		0.32
1740	Maximum	"	0.003	0.16	0.19		0.35
2520	Doors, Wood						
2540	Brush						
2560	First Coat						

PAINT

ID Code	Component Descriptions	Unit of Meas.	Manhr / Unit	Material Cost	Labor Cost	Equipment Cost	Total Cost
09 - 91131	**EXT. PAINTING, BUILDINGS, Cont'd...**						**09 - 91131**
2580	Minimum	SF	0.012	0.16	0.91		1.07
2600	Average	"	0.016	0.18	1.18		1.36
2620	Maximum	"	0.020	0.19	1.48		1.67
2640	Second Coat						
2660	Minimum	SF	0.010	0.16	0.74		0.90
2680	Average	"	0.011	0.18	0.84		1.02
2700	Maximum	"	0.013	0.19	0.98		1.17
2720	Roller						
2740	First Coat						
2760	Minimum	SF	0.005	0.16	0.39		0.55
2780	Average	"	0.007	0.18	0.49		0.67
2800	Maximum	"	0.010	0.19	0.74		0.93
2820	Second Coat						
2840	Minimum	SF	0.004	0.16	0.29		0.45
2860	Average	"	0.004	0.18	0.32		0.50
2880	Maximum	"	0.007	0.19	0.49		0.68
2900	Spray						
2920	First Coat						
2940	Minimum	SF	0.003	0.14	0.18		0.32
2960	Average	"	0.003	0.15	0.22		0.37
2980	Maximum	"	0.004	0.16	0.29		0.45
3000	Second Coat						
3020	Minimum	SF	0.002	0.14	0.14		0.28
3040	Average	"	0.002	0.15	0.16		0.31
3060	Maximum	"	0.003	0.16	0.19		0.35
3080	Gutters and Downspouts						
3100	Brush						
3120	First Coat						
3140	Minimum	LF	0.010	0.21	0.74		0.95
3160	Average	"	0.011	0.22	0.84		1.06
3180	Maximum	"	0.013	0.23	0.98		1.21
3200	Second Coat						
3220	Minimum	LF	0.007	0.21	0.49		0.70
3240	Average	"	0.008	0.22	0.59		0.81
3260	Maximum	"	0.010	0.23	0.74		0.97
3680	Siding, Wood						
3700	Roller						
3720	First Coat						

PAINT

ID Code	Description Component Descriptions	Output		Unit Costs			
		Unit of Meas.	Manhr / Unit	Material Cost	Labor Cost	Equipment Cost	Total Cost
09 - 91131	**EXT. PAINTING, BUILDINGS, Cont'd...**						**09 - 91131**
3740	Minimum	SF	0.003	0.14	0.21		0.35
3760	Average	"	0.003	0.15	0.24		0.39
3780	Maximum	"	0.004	0.16	0.26		0.42
3800	Second Coat						
3820	Minimum	SF	0.003	0.14	0.24		0.38
3840	Average	"	0.004	0.15	0.26		0.41
3860	Maximum	"	0.004	0.16	0.29		0.45
3880	Spray						
3900	First Coat						
3920	Minimum	SF	0.003	0.14	0.19		0.33
3940	Average	"	0.003	0.15	0.21		0.36
3960	Maximum	"	0.003	0.16	0.22		0.38
3980	Second Coat						
4000	Minimum	SF	0.002	0.14	0.14		0.28
4020	Average	"	0.003	0.15	0.19		0.34
4040	Maximum	"	0.004	0.16	0.29		0.45
4060	Stucco						
4080	Roller						
4100	First Coat						
4120	Minimum	SF	0.004	0.21	0.26		0.47
4140	Average	"	0.004	0.22	0.31		0.53
4160	Maximum	"	0.005	0.23	0.37		0.60
4180	Second Coat						
4200	Minimum	SF	0.003	0.21	0.21		0.42
4220	Average	"	0.003	0.22	0.25		0.47
4240	Maximum	"	0.004	0.23	0.29		0.52
4260	Spray						
4280	First Coat						
4300	Minimum	SF	0.003	0.16	0.18		0.34
4320	Average	"	0.003	0.18	0.21		0.39
4340	Maximum	"	0.003	0.19	0.24		0.43
4360	Second Coat						
4380	Minimum	SF	0.002	0.16	0.14		0.30
4400	Average	"	0.002	0.18	0.16		0.34
4420	Maximum	"	0.003	0.19	0.19		0.38
4440	Trim						
4460	Brush						
4480	First Coat						

PAINT

ID Code	Description	Output		Unit Costs			
	Component Descriptions	Unit of Meas.	Manhr / Unit	Material Cost	Labor Cost	Equipment Cost	Total Cost
09 - 91131	**EXT. PAINTING, BUILDINGS, Cont'd...**						**09 - 91131**
4500	Minimum	LF	0.003	0.21	0.24		0.45
4520	Average	"	0.004	0.22	0.29		0.51
4540	Maximum	"	0.005	0.23	0.37		0.60
4560	Second Coat						
4580	Minimum	LF	0.003	0.21	0.18		0.39
4600	Average	"	0.003	0.22	0.24		0.46
4620	Maximum	"	0.005	0.23	0.37		0.60
4640	Walls						
4660	Roller						
4680	First Coat						
4700	Minimum	SF	0.003	0.16	0.21		0.37
4720	Average	"	0.003	0.18	0.21		0.39
4740	Maximum	"	0.003	0.19	0.23		0.42
4760	Second Coat						
4780	Minimum	SF	0.003	0.16	0.18		0.34
4800	Average	"	0.003	0.18	0.19		0.37
4820	Maximum	"	0.003	0.19	0.22		0.41
4840	Spray						
4860	First Coat						
4880	Minimum	SF	0.001	0.12	0.09		0.21
4900	Average	"	0.002	0.14	0.11		0.25
4920	Maximum	"	0.002	0.15	0.14		0.29
4940	Second Coat						
4960	Minimum	SF	0.001	0.12	0.07		0.19
4980	Average	"	0.001	0.14	0.09		0.23
5000	Maximum	"	0.002	0.15	0.13		0.28
5020	Windows						
5040	Brush						
5060	First Coat						
5080	Minimum	SF	0.013	0.14	0.98		1.12
5100	Average	"	0.016	0.15	1.18		1.33
5120	Maximum	"	0.020	0.16	1.48		1.64
5140	Second Coat						
5160	Minimum	SF	0.011	0.14	0.84		0.98
5180	Average	"	0.013	0.15	0.98		1.13
5200	Maximum	"	0.016	0.16	1.18		1.34

PAINT

ID Code	Description		Output		Unit Costs			
	Component Descriptions		Unit of Meas.	Manhr / Unit	Material Cost	Labor Cost	Equipment Cost	Total Cost
09 - 91132	**EXT. PAINTING, MISC.**							**09 - 91132**
3000	Shakes							
3020	Spray							
3040	First Coat							
3060	Minimum		SF	0.003	0.15	0.24		0.39
3080	Average		"	0.004	0.16	0.26		0.42
3100	Maximum		"	0.004	0.18	0.29		0.47
3120	Second Coat							
3140	Minimum		SF	0.003	0.15	0.22		0.37
3160	Average		"	0.003	0.16	0.24		0.40
3180	Maximum		"	0.004	0.18	0.26		0.44
3200	Shingles, Wood							
3220	Roller							
3240	First Coat							
3260	Minimum		SF	0.004	0.16	0.32		0.48
3280	Average		"	0.005	0.18	0.37		0.55
3300	Maximum		"	0.006	0.19	0.42		0.61
3320	Second Coat							
3340	Minimum		SF	0.003	0.16	0.22		0.38
3360	Average		"	0.003	0.18	0.24		0.42
3380	Maximum		"	0.004	0.19	0.26		0.45
3400	Spray							
3420	First Coat							
3440	Minimum		LF	0.003	0.14	0.22		0.36
3460	Average		"	0.003	0.15	0.24		0.39
3480	Maximum		"	0.004	0.16	0.26		0.42
3500	Second Coat							
3520	Minimum		LF	0.002	0.14	0.17		0.31
3540	Average		"	0.003	0.15	0.18		0.33
3560	Maximum		"	0.003	0.16	0.19		0.35
4000	Shutters and Louvers							
4020	Brush							
4040	First Coat							
4060	Minimum		EA	0.160	0.21	11.75		12.00
4080	Average		"	0.200	0.22	14.75		15.00
4100	Maximum		"	0.267	0.23	19.75		20.00
4120	Second Coat							
4140	Minimum		EA	0.100	0.21	7.41		7.62
4160	Average		"	0.123	0.22	9.12		9.34

PAINT

ID Code	Description — Component Descriptions	Output — Unit of Meas.	Output — Manhr / Unit	Unit Costs — Material Cost	Unit Costs — Labor Cost	Unit Costs — Equipment Cost	Unit Costs — Total Cost
09 - 91132	**EXT. PAINTING, MISC., Cont'd...**						**09 - 91132**
4180	Maximum	EA	0.160	0.23	11.75		12.00
4200	Spray						
4220	First Coat						
4240	Minimum	EA	0.053	0.15	3.95		4.10
4260	Average	"	0.064	0.16	4.74		4.90
4280	Maximum	"	0.080	0.18	5.92		6.10
4300	Second Coat						
4320	Minimum	EA	0.040	0.15	2.96		3.11
4340	Average	"	0.053	0.16	3.95		4.11
4360	Maximum	"	0.064	0.18	4.74		4.92
5000	Stairs, metal						
5020	Brush						
5040	First Coat						
5060	Minimum	SF	0.009	0.21	0.65		0.86
5080	Average	"	0.010	0.22	0.74		0.96
5100	Maximum	"	0.011	0.23	0.84		1.07
5120	Second Coat						
5140	Minimum	SF	0.005	0.21	0.37		0.58
5160	Average	"	0.006	0.22	0.42		0.64
5180	Maximum	"	0.007	0.23	0.49		0.72
5200	Spray						
5220	First Coat						
5240	Minimum	SF	0.004	0.15	0.32		0.47
5260	Average	"	0.006	0.16	0.42		0.58
5280	Maximum	"	0.006	0.18	0.45		0.63
5300	Second Coat						
5320	Minimum	SF	0.003	0.15	0.24		0.39
5340	Average	"	0.004	0.16	0.29		0.45
5360	Maximum	"	0.005	0.18	0.37		0.55
09 - 91233	**INT. PAINTING, BUILDINGS**						**09 - 91233**
1000	Acoustical Ceiling						
1020	Roller						
1040	First Coat						
1060	Minimum	SF	0.005	0.21	0.37		0.58
1080	Average	"	0.007	0.22	0.49		0.71
1100	Maximum	"	0.010	0.23	0.74		0.97
1120	Second Coat						

PAINT

ID Code	Description		Output		Unit Costs			
	Component Descriptions		Unit of Meas.	Manhr / Unit	Material Cost	Labor Cost	Equipment Cost	Total Cost
09 - 91233	**INT. PAINTING, BUILDINGS, Cont'd...**							**09 - 91233**
1140	Minimum		SF	0.004	0.21	0.29		0.50
1160	Average		"	0.005	0.22	0.37		0.59
1180	Maximum		"	0.007	0.23	0.49		0.72
1200	Spray							
1220	First Coat							
1240	Minimum		SF	0.002	0.16	0.16		0.32
1260	Average		"	0.003	0.18	0.19		0.37
1280	Maximum		"	0.003	0.19	0.24		0.43
1300	Second Coat							
1320	Minimum		SF	0.002	0.16	0.13		0.29
1340	Average		"	0.002	0.18	0.14		0.32
1360	Maximum		"	0.002	0.19	0.16		0.35
1380	Cabinets and Casework							
1400	Brush							
1420	First Coat							
1440	Minimum		SF	0.008	0.21	0.59		0.80
1460	Average		"	0.009	0.22	0.65		0.87
1480	Maximum		"	0.010	0.23	0.74		0.97
1500	Second Coat							
1520	Minimum		SF	0.007	0.21	0.49		0.70
1540	Average		"	0.007	0.22	0.53		0.75
1560	Maximum		"	0.008	0.23	0.59		0.82
1580	Spray							
1600	First Coat							
1620	Minimum		SF	0.004	0.16	0.29		0.45
1640	Average		"	0.005	0.18	0.34		0.52
1660	Maximum		"	0.006	0.19	0.42		0.61
1680	Second Coat							
1700	Minimum		SF	0.003	0.16	0.23		0.39
1720	Average		"	0.003	0.18	0.25		0.43
1740	Maximum		"	0.004	0.19	0.32		0.51
1760	Ceilings							
1780	Roller							
1800	First Coat							
1820	Minimum		SF	0.003	0.16	0.24		0.40
1840	Average		"	0.004	0.18	0.26		0.44
1860	Maximum		"	0.004	0.19	0.29		0.48
1880	Second Coat							

PAINT

ID Code	Description — Component Descriptions	Output — Unit of Meas.	Output — Manhr / Unit	Unit Costs — Material Cost	Unit Costs — Labor Cost	Unit Costs — Equipment Cost	Unit Costs — Total Cost
09 - 91233	**INT. PAINTING, BUILDINGS, Cont'd...**						**09 - 91233**
1900	Minimum	SF	0.003	0.16	0.19		0.35
1920	Average	"	0.003	0.18	0.22		0.40
1940	Maximum	"	0.003	0.19	0.24		0.43
1960	Spray						
1980	First Coat						
2000	Minimum	SF	0.002	0.14	0.14		0.28
2020	Average	"	0.002	0.15	0.16		0.31
2040	Maximum	"	0.003	0.16	0.18		0.34
2060	Second Coat						
2080	Minimum	SF	0.002	0.14	0.11		0.25
2100	Average	"	0.002	0.15	0.12		0.27
2120	Maximum	"	0.002	0.16	0.14		0.30
2520	Doors, Wood						
2540	Brush						
2560	First Coat						
2580	Minimum	SF	0.011	0.21	0.84		1.05
2600	Average	"	0.015	0.22	1.07		1.29
2620	Maximum	"	0.018	0.23	1.31		1.54
2640	Second Coat						
2660	Minimum	SF	0.009	0.15	0.65		0.80
2680	Average	"	0.010	0.16	0.74		0.90
2700	Maximum	"	0.011	0.18	0.84		1.02
2720	Spray						
2740	First Coat						
2760	Minimum	SF	0.002	0.15	0.17		0.32
2780	Average	"	0.003	0.16	0.21		0.37
2800	Maximum	"	0.004	0.18	0.26		0.44
2820	Second Coat						
2840	Minimum	SF	0.002	0.15	0.14		0.29
2860	Average	"	0.002	0.16	0.16		0.32
2880	Maximum	"	0.003	0.18	0.18		0.36
3900	Trim						
3920	Brush						
3940	First Coat						
3960	Minimum	LF	0.003	0.21	0.23		0.44
3980	Average	"	0.004	0.22	0.26		0.48
4000	Maximum	"	0.004	0.23	0.32		0.55
4020	Second Coat						

PAINT

ID Code	Component Descriptions	Unit of Meas.	Manhr / Unit	Material Cost	Labor Cost	Equipment Cost	Total Cost
	Description			**Unit Costs**			
09 - 91233	**INT. PAINTING, BUILDINGS, Cont'd...**						**09 - 91233**
4040	Minimum	LF	0.002	0.21	0.17		0.38
4060	Average	"	0.003	0.22	0.22		0.44
4080	Maximum	"	0.004	0.23	0.32		0.55
4100	Walls						
4120	Roller						
4140	First Coat						
4160	Minimum	SF	0.003	0.16	0.21		0.37
4180	Average	"	0.003	0.18	0.21		0.39
4200	Maximum	"	0.003	0.19	0.24		0.43
4220	Second Coat						
4240	Minimum	SF	0.003	0.16	0.18		0.34
4260	Average	"	0.003	0.18	0.19		0.37
4280	Maximum	"	0.003	0.19	0.22		0.41
4300	Spray						
4320	First Coat						
4340	Minimum	SF	0.001	0.14	0.09		0.23
4360	Average	"	0.002	0.15	0.11		0.26
4380	Maximum	"	0.002	0.16	0.14		0.30
4400	Second Coat						
4420	Minimum	SF	0.001	0.14	0.08		0.22
4440	Average	"	0.001	0.15	0.10		0.25
4460	Maximum	"	0.002	0.16	0.13		0.29

DIVISION 10
SPECIALTIES

COMPARTMENTS & CUBICLES

ID Code	Component Descriptions	Unit of Meas.	Manhr / Unit	Material Cost	Labor Cost	Equipment Cost	Total Cost
		Output		**Unit Costs**			
	Description						

10 - 21000 **SHOWER STALLS** **10 - 21000**

ID Code	Component Descriptions	Unit of Meas.	Manhr / Unit	Material Cost	Labor Cost	Equipment Cost	Total Cost
1000	Shower receptors						
1010	Precast, terrazzo						
1020	32" x 32"	EA	0.667	850	65.00		920
1040	32" x 48"	"	0.800	900	78.00		980
1050	Concrete						
1060	32" x 32"	EA	0.667	350	65.00		420
1080	48" x 48"	"	0.889	390	87.00		480
1100	Shower door, trim and hardware						
1120	Economy, 24" wide, chrome, tempered glass	EA	0.800	380	78.00		460
1130	Porcelain enameled steel, flush	"	0.800	700	78.00		780
1140	Baked enameled steel, flush	"	0.800	410	78.00		490
1150	Aluminum, tempered glass, 48" wide, sliding	"	1.000	860	97.00		960
1161	Folding	"	1.000	830	97.00		930
1190	Aluminum and tempered glass, molded plastic						
1200	Complete with receptor and door						
1220	32" x 32"	EA	2.000	1,050	190		1,240
1230	36" x 36"	"	2.000	1,180	190		1,370
1240	40" x 40"	"	2.286	1,230	220		1,450

TOILET, BATH AND LAUNDRY ACCESSORIES

10 - 28160 **BATH ACCESSORIES** **10 - 28160**

ID Code	Component Descriptions	Unit of Meas.	Manhr / Unit	Material Cost	Labor Cost	Equipment Cost	Total Cost
1050	Grab bar, 1-1/2" dia., stainless steel, wall mounted						
1060	24" long	EA	0.400	91.00	35.50		130
1080	36" long	"	0.421	100	37.50		140
1130	1" dia., stainless steel						
1140	12" long	EA	0.348	57.00	31.00		88.00
1180	24" long	"	0.400	77.00	35.50		110
1220	36" long	"	0.444	100	39.50		140
1320	Medicine cabinet, 16 x 22, baked enamel, lighted	"	0.320	260	28.50		290
1340	With mirror, lighted	"	0.533	380	47.25		430
1420	Mirror, 1/4" plate glass, up to 10 sf	SF	0.080	20.00	7.10		27.00
1430	Mirror, stainless steel frame						
1440	18"x24"	EA	0.267	150	23.75		170
1460	18"x32"	"	0.320	180	28.50		210
1500	24"x30"	"	0.400	190	35.50		230
1530	24"x60"	"	0.800	700	71.00		770
1900	Soap dish, stainless steel, wall mounted	"	0.533	260	47.25		310

TOILET, BATH AND LAUNDRY ACCESSORIES

ID Code	Description Component Descriptions	Output		Unit Costs			
		Unit of Meas.	Manhr / Unit	Material Cost	Labor Cost	Equipment Cost	Total Cost
10 - 28160	**BATH ACCESSORIES, Cont'd...**						**10 - 28160**
1910	Toilet tissue dispenser, stainless, wall mounted						
1920	Single roll	EA	0.200	130	17.75		150
2000	Towel bar, stainless steel						
2020	18" long	EA	0.320	160	28.50		190
2040	24" long	"	0.364	220	32.25		250
2060	30" long	"	0.400	220	35.50		260
2070	36" long	"	0.444	240	39.50		280
2080	Toothbrush and tumbler holder	"	0.267	99.00	23.75		120

FLAGPOLES

ID Code	Component Descriptions	Unit of Meas.	Manhr / Unit	Material Cost	Labor Cost	Equipment Cost	Total Cost
10 - 75001	**FLAGPOLES**						**10 - 75001**
2020	Installed in concrete base						
2030	Fiberglass						
2040	25' high	EA	5.333	2,210	470		2,680
2080	50' high	"	13.333	5,840	1,180		7,020
2100	Aluminum						
2120	25' high	EA	5.333	2,140	470		2,610
2140	50' high	"	13.333	4,240	1,180		5,420
2160	Bonderized steel						
2180	25' high	EA	6.154	2,400	550		2,950
2200	50' high	"	16.000	4,800	1,420		6,220
2220	Freestanding tapered, fiberglass						
2240	30' high	EA	5.714	2,630	510		3,140
2260	40' high	"	7.273	3,410	650		4,060

PEST CONTROL DEVICES

ID Code	Component Descriptions	Unit of Meas.	Manhr / Unit	Material Cost	Labor Cost	Equipment Cost	Total Cost
10 - 81001	**PEST CONTROL**						**10 - 81001**
1000	Termite control						
1010	Under slab spraying						
1020	Minimum	SF	0.002	1.31	0.13		1.44
1040	Average	"	0.004	1.31	0.27		1.58
1120	Maximum	"	0.008	1.87	0.55		2.42

DIVISION 11
EQUIPMENT

ARCHITECTURAL EQUIPMENT

ID Code	Component Descriptions	Unit of Meas.	Manhr / Unit	Material Cost	Labor Cost	Equipment Cost	Total Cost
	Description	**Output**		**Unit Costs**			
11 - 16001	**VAULTS**					**11 - 16001**	
1000	Floor safes						
1010	1.0 cf	EA	0.667	1,300	59.00		1,360
1020	1.3 cf	"	1.000	1,440	89.00		1,530
11 - 24001	**MAINTENANCE EQUIPMENT**					**11 - 24001**	
1000	Vacuum cleaning system						
1010	3 valves						
1020	1.5 hp	EA	8.889	1,180	790		1,970
1030	2.5 hp	"	11.429	1,420	1,010		2,430
1040	5 valves	"	16.000	2,220	1,420		3,640
1060	7 valves	"	20.000	2,950	1,780		4,730
11 - 31001	**RESIDENTIAL EQUIPMENT**					**11 - 31001**	
0300	Compactor, 4 to 1 compaction	EA	2.000	2,350	180		2,530
1300	Dishwasher, built-in						
1320	2 cycles	EA	4.000	1,160	360		1,520
1330	4 or more cycles	"	4.000	3,120	360		3,480
1340	Disposal						
1350	Garbage disposer	EA	2.667	320	240		560
1360	Heaters, electric, built-in						
1362	Ceiling type	EA	2.667	660	240		900
1363	Wall type						
1370	Minimum	EA	2.000	330	180		510
1380	Maximum	"	2.667	1,150	240		1,390
1390	Hood for range, 2-speed, vented						
1420	30" wide	EA	2.667	910	240		1,150
1440	42" wide	"	2.667	1,680	240		1,920
1460	Ice maker, automatic						
1480	30 lb per day	EA	1.143	3,080	100		3,180
1500	50 lb per day	"	4.000	3,900	360		4,260
1820	Folding access stairs, disappearing metal stair						
1840	8' long	EA	1.143	1,610	100		1,710
1850	11' long	"	1.143	1,680	100		1,780
1860	12' long	"	1.143	1,790	100		1,890
1940	Wood frame, wood stair						
1950	22" x 54" x 8'9" long	EA	0.800	310	72.00		380
1960	25" x 54" x 10' long	"	0.800	380	72.00		450
2020	Ranges, electric						
2040	Built-in, 30", 1 oven	EA	2.667	3,370	240		3,610

ARCHITECTURAL EQUIPMENT

ID Code	Component Descriptions	Unit of Meas.	Manhr / Unit	Material Cost	Labor Cost	Equipment Cost	Total Cost
		Description		**Output**		**Unit Costs**	
11 - 31001	**RESIDENTIAL EQUIPMENT, Cont'd...**					**11 - 31001**	
2050	2 oven	EA	2.667	3,900	240		4,140
2060	Countertop, 4 burner, standard	"	2.000	1,950	180		2,130
2070	With grill	"	2.000	4,870	180		5,050
2198	Freestanding, 21", 1 oven	"	2.667	1,760	240		2,000
2200	30", 1 oven	"	1.600	3,410	140		3,550
2220	2 oven	"	1.600	5,560	140		5,700
3600	Water softener						
3620	30 grains per gallon	EA	2.667	1,910	240		2,150
3640	70 grains per gallon	"	4.000	2,400	360		2,760
11 - 68230	**RECREATIONAL COURTS**					**11 - 68230**	
1000	Walls, galvanized steel						
1020	8' high	LF	0.160	17.50	11.25		28.75
1040	10' high	"	0.178	20.75	12.25		33.00
1060	12' high	"	0.211	23.75	14.75		38.50
1200	Vinyl coated						
1220	8' high	LF	0.160	16.75	11.25		28.00
1240	10' high	"	0.178	20.50	12.25		32.75
1260	12' high	"	0.211	23.00	14.75		37.75
2010	Gates, galvanized steel						
2200	Single, 3' transom						
2210	3'x7'	EA	4.000	510	280		790
2220	4'x7'	"	4.571	540	320		860
2230	5'x7'	"	5.333	740	370		1,110
2240	6'x7'	"	6.400	800	450		1,250
2400	Vinyl coated						
2405	Single, 3' transom						
2410	3'x7'	EA	4.000	990	280		1,270
2420	4'x7'	"	4.571	1,080	320		1,400
2430	5'x7'	"	5.333	1,080	370		1,450
2440	6'x7'	"	6.400	1,110	450		1,560

RECYCLING SYSTEMS

ID Code	Description Component Descriptions	Output Unit of Meas.	Manhr / Unit	Unit Costs Material Cost	Labor Cost	Equipment Cost	Total Cost
11 - 82230	**GRAY WATER RECYCLING SYSTEM**						**11 - 82230**
1000	Residential, small commercial, 150 Gallons						
1010	Minimum	EA					5,030
1020	Average	"					5,700
1030	Maximum	"					6,370
1040	250 Gallons						
1050	Minimum	EA					5,860
1060	Average	"					6,530
1070	Maximum	"					7,200
1080	350 Gallons						
1090	Minimum	EA					6,370
1100	Average	"					6,870
1110	Maximum	"					7,370
1120	450 Gallons						
1130	Minimum	EA					6,700
1140	Average	"					7,200
1150	Maximum	"					7,710
1160	550 Gallons						
1170	Minimum	EA					8,210
1180	Average	"					6,790
1190	Maximum	"					8,710

DIVISION 12
FURNISHINGS

WINDOW BLINDS

ID Code	Description / Component Descriptions	Output / Unit of Meas.	Manhr / Unit	Material Cost	Labor Cost	Equipment Cost	Total Cost
12 - 21001	**BLINDS**						**12 - 21001**
0990	Venetian blinds						
1000	2" slats	SF	0.020	48.25	1.77		50.00
1020	1" slats	"	0.020	52.00	1.77		54.00

CURTAINS AND DRAPES

ID Code	Component Descriptions	Unit of Meas.	Manhr / Unit	Material Cost	Labor Cost	Equipment Cost	Total Cost
12 - 22001	**WINDOW TREATMENT**						**12 - 22001**
1000	Drapery tracks, wall or ceiling mounted						
1040	Basic traverse rod						
1080	50 to 90"	EA	0.400	64.00	35.50		100
1100	84 to 156"	"	0.444	85.00	39.50		120
1120	136 to 250"	"	0.444	120	39.50		160
1140	165 to 312"	"	0.500	190	44.50		230
1160	Traverse rod with stationary curtain rod						
1180	30 to 50"	EA	0.400	96.00	35.50		130
1200	50 to 90"	"	0.400	110	35.50		150
1220	84 to 156"	"	0.444	150	39.50		190
1240	136 to 250"	"	0.500	190	44.50		230
1260	Double traverse rod						
1280	30 to 50"	EA	0.400	110	35.50		150
1300	50 to 84"	"	0.400	140	35.50		180
1320	84 to 156"	"	0.444	150	39.50		190
1340	136 to 250"	"	0.500	200	44.50		240

MANUFACTURED WOOD CASEWORK

ID Code	Component Descriptions	Unit of Meas.	Manhr / Unit	Material Cost	Labor Cost	Equipment Cost	Total Cost
12 - 32001	**CASEWORK**						**12 - 32001**
0080	Kitchen base cabinet, standard, 24" deep, 35" high						
0100	12" wide	EA	0.800	250	71.00		320
0120	18" wide	"	0.800	290	71.00		360
0140	24" wide	"	0.889	370	79.00		450
0160	27" wide	"	0.889	420	79.00		500
0180	36" wide	"	1.000	500	89.00		590
0200	48" wide	"	1.000	600	89.00		690
0210	Drawer base, 24" deep, 35" high						
0220	15" wide	EA	0.800	310	71.00		380
0230	18" wide	"	0.800	330	71.00		400
0240	24" wide	"	0.889	540	79.00		620

MANUFACTURED WOOD CASEWORK

ID Code	Description		Output		Unit Costs			
	Component Descriptions	Unit of Meas.	Manhr / Unit	Material Cost	Labor Cost	Equipment Cost	Total Cost
12 - 32001	**CASEWORK, Cont'd...**						**12 - 32001**
0250	27" wide	EA	0.889	610	79.00		690
0260	30" wide	"	0.889	710	79.00		790
0270	Sink-ready base cabinet						
0280	30" wide	EA	0.889	330	79.00		410
0290	36" wide	"	0.889	350	79.00		430
0300	42" wide	"	0.889	380	79.00		460
0310	60" wide	"	1.000	450	89.00		540
0320	Corner cabinet, 36" wide	"	1.000	630	89.00		720
4000	Wall cabinet, 12" deep, 12" high						
4020	30" wide	EA	0.800	320	71.00		390
4060	36" wide	"	0.800	330	71.00		400
4070	15" high						
4080	30" wide	EA	0.889	370	79.00		450
4100	36" wide	"	0.889	560	79.00		640
4110	24" high						
4120	30" wide	EA	0.889	410	79.00		490
4140	36" wide	"	0.889	430	79.00		510
4150	30" high						
4160	12" wide	EA	1.000	240	89.00		330
4180	18" wide	"	1.000	270	89.00		360
4200	24" wide	"	1.000	290	89.00		380
4300	27" wide	"	1.000	350	89.00		440
4320	30" wide	"	1.143	390	100		490
4340	36" wide	"	1.143	400	100		500
4350	Corner cabinet, 30" high						
4360	24" wide	EA	1.333	440	120		560
4380	30" wide	"	1.333	530	120		650
4390	36" wide	"	1.333	580	120		700
5020	Wardrobe	"	2.000	1,160	180		1,340
6980	Vanity with top, laminated plastic						
7000	24" wide	EA	2.000	960	180		1,140
7020	30" wide	"	2.000	1,070	180		1,250
7040	36" wide	"	2.667	1,240	240		1,480
7060	48" wide	"	3.200	1,380	280		1,660

COUNTERTOPS

ID Code	Description	Output		Unit Costs			
	Component Descriptions	Unit of Meas.	Manhr / Unit	Material Cost	Labor Cost	Equipment Cost	Total Cost
12 - 36001	**COUNTERTOPS**						**12 - 36001**
1020	Stainless steel, countertop, with backsplash	SF	0.200	290	17.75		310
2000	Acid-proof, kemrock surface	"	0.133	120	11.75		130

DIVISION 13
SPECIAL CONSTRUCTION

CONSTRUCTION

ID Code	Description	Output		Unit Costs			
	Component Descriptions	Unit of Meas.	Manhr / Unit	Material Cost	Labor Cost	Equipment Cost	Total Cost
13 - 34190	**PRE-ENGINEERED BUILDINGS**						**13 - 34190**
1080	Pre-engineered metal building, 40'x100'						
1100	14' eave height	SF	0.032	11.75	2.84	3.66	18.25
1120	16' eave height	"	0.037	13.50	3.27	4.23	21.00

DIVISION 14
CONVEYING

LIFTS

ID Code	Component Descriptions	Unit of Meas.	Manhr / Unit	Material Cost	Labor Cost	Equipment Cost	Total Cost
	Description	**Output**		**Unit Costs**			
14 - 41001		**PERSONNEL LIFTS**					**14 - 41001**
1000	Electrically operated, 1 or 2 person lift						
1001	With attached foot platforms						
1020	3 stops	EA					13,250
3020	Residential stair climber, per story	"	6.667	6,140	600		6,740
14 - 42001		**WHEELCHAIR LIFTS**					**14 - 42001**
1000	600 lb, Residential	EA	8.000	7,300	720		8,020

DIVISION 21
FIRE SUPPRESSION

COMPONENTS

ID Code	Component Descriptions	Unit of Meas.	Manhr / Unit	Material Cost	Labor Cost	Equipment Cost	Total Cost
	Description	**Output**		**Unit Costs**			
21 - 11160		**HYDRANTS**					**21 - 11160**
0980	Wall hydrant						
1000	8" thick	EA	1.333	400	130		530
1020	12" thick	"	1.600	470	160		630

DIVISION 22
PLUMBING

BASIC MATERIALS

ID Code	Component Descriptions	Unit of Meas.	Manhr / Unit	Material Cost	Labor Cost	Equipment Cost	Total Cost
	Description						

ID Code	Component Descriptions	Unit of Meas.	Manhr / Unit	Material Cost	Labor Cost	Equipment Cost	Total Cost
22 - 05236	**VALVES**						**22 - 05236**
0600	Gate valve, 125 lb, bronze, soldered						
0800	1/2"	EA	0.200	43.75	19.50		63.00
1000	3/4"	"	0.200	52.00	19.50		72.00
1055	Threaded						
1058	1/4", 125 lb	EA	0.320	37.50	31.25		69.00
1059	1/2"						
1060	125 lb	EA	0.320	35.75	31.25		67.00
1070	150 lb	"	0.320	48.00	31.25		79.00
1075	300 lb	"	0.320	90.00	31.25		120
1078	3/4"						
1083	125 lb	EA	0.320	42.00	31.25		73.00
1086	150 lb	"	0.320	57.00	31.25		88.00
1088	300 lb	"	0.320	110	31.25		140
1280	Check valve, bronze, soldered, 125 lb						
1300	1/2"	EA	0.200	65.00	19.50		85.00
1320	3/4"	"	0.200	81.00	19.50		100
1365	Threaded						
1367	1/2"						
1370	125 lb	EA	0.267	76.00	26.00		100
1380	150 lb	"	0.267	70.00	26.00		96.00
1390	200 lb	"	0.267	73.00	26.00		99.00
1395	3/4"						
1400	125 lb	EA	0.320	56.00	31.25		87.00
1410	150 lb	"	0.320	88.00	31.25		120
1420	200 lb	"	0.320	97.00	31.25		130
1671	Vertical check valve, bronze, 125 lb, threaded						
1673	1/2"	EA	0.320	86.00	31.25		120
1674	3/4"	"	0.364	130	35.50		170
1790	Globe valve, bronze, soldered, 125 lb						
1800	1/2"	EA	0.229	96.00	22.25		120
1810	3/4"	"	0.250	120	24.25		140
1980	Threaded						
1990	1/2"						
2000	125 lb	EA	0.267	77.00	26.00		100
2020	150 lb	"	0.267	100	26.00		130
2030	300 lb	"	0.267	190	26.00		220
2035	3/4"						
2040	125 lb	EA	0.320	110	31.25		140

BASIC MATERIALS

ID Code	Description — Component Descriptions	Output — Unit of Meas.	Output — Manhr / Unit	Unit Costs — Material Cost	Unit Costs — Labor Cost	Unit Costs — Equipment Cost	Unit Costs — Total Cost
22 - 05236	**VALVES, Cont'd...**						**22 - 05236**
2060	150 lb	EA	0.320	120	31.25		150
2070	300 lb	"	0.320	230	31.25		260
3980	Ball valve, bronze, 250 lb, threaded						
4000	1/2"	EA	0.320	20.50	31.25		52.00
4010	3/4"	"	0.320	30.50	31.25		62.00
4980	Angle valve, bronze, 150 lb, threaded						
5000	1/2"	EA	0.286	100	27.75		130
5010	3/4"	"	0.320	140	31.25		170
5980	Balancing valve, meter connections, circuit setter						
6000	1/2"	EA	0.320	99.00	31.25		130
6010	3/4"	"	0.364	100	35.50		140
6090	Balancing valve, straight type						
6100	1/2"	EA	0.320	48.25	31.25		80.00
6120	3/4"	"	0.320	59.00	31.25		90.00
6125	Angle type						
6130	1/2"	EA	0.320	65.00	31.25		96.00
6140	3/4"	"	0.320	91.00	31.25		120
6580	Square head cock, 125 lb, bronze body						
6600	1/2"	EA	0.267	21.25	26.00		47.25
6610	3/4"	"	0.320	25.50	31.25		57.00
6635	Radiator temp control valve, with control and sensor						
6640	1/2" valve	EA	0.500	150	48.75		200
7980	Pressure relief valve, 1/2", bronze						
8000	Low pressure	EA	0.320	34.00	31.25		65.00
8020	High pressure	"	0.320	39.50	31.25		71.00
8030	Pressure and temperature relief valve						
8040	Bronze, 3/4"	EA	0.320	180	31.25		210
8045	Cast iron, 3/4"						
8050	High pressure	EA	0.320	88.00	31.25		120
8060	Temperature relief	"	0.320	120	31.25		150
8070	Pressure & temp relief valve	"	0.320	140	31.25		170
8100	Pressure reducing valve, bronze, threaded, 250 lb						
8120	1/2"	EA	0.500	200	48.75		250
8140	3/4"	"	0.500	200	48.75		250
8480	Solar water temperature regulating valve						
8500	3/4"	EA	0.667	770	65.00		830
8980	Tempering valve, threaded						
9000	3/4"	EA	0.267	470	26.00		500

BASIC MATERIALS

ID Code	Description / Component Descriptions	Output Unit of Meas.	Manhr / Unit	Material Cost	Labor Cost	Equipment Cost	Total Cost
22 - 05236	**VALVES, Cont'd...**						**22 - 05236**
9180	Thermostatic mixing valve, threaded						
9200	1/2"	EA	0.286	130	27.75		160
9210	3/4"	"	0.320	130	31.25		160
9245	Sweat connection						
9250	1/2"	EA	0.286	110	27.75		140
9260	3/4"	"	0.320	140	31.25		170
9265	Mixing valve, sweat connection						
9270	1/2"	EA	0.286	79.00	27.75		110
9280	3/4"	"	0.320	79.00	31.25		110
9480	Liquid level gauge, aluminum body						
9500	3/4"	EA	0.320	400	31.25		430
9505	125 psi, PVC body						
9510	3/4"	EA	0.320	470	31.25		500
9520	150 psi, CRS body						
9530	3/4"	EA	0.320	380	31.25		410
9560	175 psi, bronze body, 1/2"	"	0.286	760	27.75		790
22 - 06291	**PIPE HANGERS, LIGHT**						**22 - 06291**
0010	A band, black iron						
0020	1/2"	EA	0.057	1.03	5.56		6.59
0030	1"	"	0.059	1.11	5.77		6.88
0040	1-1/4"	"	0.062	1.23	5.99		7.22
0050	1-1/2"	"	0.067	1.28	6.49		7.77
0060	2"	"	0.073	1.36	7.08		8.44
0070	2-1/2"	"	0.080	2.03	7.79		9.82
0080	3"	"	0.089	2.48	8.65		11.25
0090	4"	"	0.100	3.26	9.74		13.00
0130	Copper						
0140	1/2"	EA	0.057	1.67	5.56		7.23
0150	3/4"	"	0.059	1.94	5.77		7.71
0160	1"	"	0.059	1.94	5.77		7.71
0170	1-1/4"	"	0.062	2.09	5.99		8.08
0180	1-1/2"	"	0.067	2.24	6.49		8.73
0190	2"	"	0.073	2.37	7.08		9.45
0200	2-1/2"	"	0.080	4.79	7.79		12.50
0210	3"	"	0.089	4.99	8.65		13.75
0220	4"	"	0.100	5.51	9.74		15.25
1000	2 hole clips, galvanized						

BASIC MATERIALS

ID Code	Description / Component Descriptions	Output Unit of Meas.	Output Manhr / Unit	Unit Costs Material Cost	Unit Costs Labor Cost	Unit Costs Equipment Cost	Unit Costs Total Cost
22 - 06291	**PIPE HANGERS, LIGHT, Cont'd...**						**22 - 06291**
1030	3/4"	EA	0.053	0.27	5.19		5.46
1040	1"	"	0.055	0.33	5.37		5.70
1050	1-1/4"	"	0.057	0.43	5.56		5.99
1060	1-1/2"	"	0.059	0.53	5.77		6.30
1070	2"	"	0.062	0.70	5.99		6.69
1080	2-1/2"	"	0.064	1.25	6.23		7.48
1090	3"	"	0.067	1.66	6.49		8.15
1110	4"	"	0.073	3.56	7.08		10.75
1120	Perforated strap						
1130	3/4"						
1140	Galvanized, 20 ga.	LF	0.040	0.44	3.89		4.33
1150	Copper, 22 ga.	"	0.040	2.20	3.89		6.09
1740	J-Hooks						
1750	1/2"	EA	0.036	0.87	3.54		4.41
1760	3/4"	"	0.036	0.93	3.54		4.47
1770	1"	"	0.038	0.95	3.71		4.66
1780	1-1/4"	"	0.039	1.00	3.80		4.80
1790	1-1/2"	"	0.040	1.02	3.89		4.91
1800	2"	"	0.040	1.07	3.89		4.96
1810	3"	"	0.042	1.23	4.10		5.33
1820	4"	"	0.042	1.33	4.10		5.43
1830	PVC coated hangers, galvanized, 28 ga.						
1840	1-1/2" x 12"	EA	0.053	1.86	5.19		7.05
1850	2" x 12"	"	0.057	2.04	5.56		7.60
1860	3" x 12"	"	0.062	2.28	5.99		8.27
1870	4" x 12"	"	0.067	2.53	6.49		9.02
1880	Copper, 30 ga.						
1890	1-1/2" x 12"	EA	0.053	2.28	5.19		7.47
1900	2" x 12"	"	0.057	2.71	5.56		8.27
1910	3" x 12"	"	0.062	3.01	5.99		9.00
1920	4" x 12"	"	0.067	3.30	6.49		9.79
2090	Wire hook hangers						
2095	Black wire, 1/2" x						
2100	4"	EA	0.040	0.44	3.89		4.33
2110	6"	"	0.042	0.50	4.10		4.60
4000	Copper wire hooks						
4010	1/2" x						
4020	4"	EA	0.040	0.58	3.89		4.47

BASIC MATERIALS

ID Code	Description / Component Descriptions	Unit of Meas.	Manhr / Unit	Material Cost	Labor Cost	Equipment Cost	Total Cost
22 - 06291	**PIPE HANGERS, LIGHT, Cont'd...**						**22 - 06291**
4030	6"	EA	0.042	0.66	4.10		4.76
22 - 06481	**VIBRATION CONTROL**						**22 - 06481**
0120	Vibration isolator, in-line, stainless connector						
0140	1/2"	EA	0.444	100	43.25		140
0160	3/4"	"	0.471	120	45.75		170
0180	1"	"	0.500	120	48.75		170
0200	1-1/4"	"	0.533	170	52.00		220
0260	1-1/2"	"	0.571	190	56.00		250
0280	2"	"	0.615	230	60.00		290
0290	2-1/2"	"	0.667	340	65.00		400
0300	3"	"	0.727	400	71.00		470
0320	4"	"	0.800	510	78.00		590
22 - 06931	**SPECIALTIES**						**22 - 06931**
1000	Wall penetration						
1010	Concrete wall, 6" thick						
1020	2" dia.	EA	0.267		18.50		18.50
1040	4" dia.	"	0.400		27.75		27.75
1090	12" thick						
1100	2" dia.	EA	0.364		25.25		25.25
1120	4" dia.	"	0.571		39.75		39.75

FACILITY WATER DISTRIBUTION

ID Code	Description / Component Descriptions	Unit of Meas.	Manhr / Unit	Material Cost	Labor Cost	Equipment Cost	Total Cost
22 - 11161	**COPPER PIPE**						**22 - 11161**
0600	Type K copper						
0890	1/4"	LF	0.024	2.12	2.29		4.41
0895	3/8"	"	0.024	3.25	2.29		5.54
0900	1/2"	"	0.025	3.77	2.43		6.20
1000	3/4"	"	0.027	7.03	2.59		9.62
1020	1"	"	0.029	9.20	2.78		12.00
1320	3-1/2"	"	0.043	61.00	4.21		65.00
3000	DWV, copper						
3020	1-1/4"	LF	0.033	10.25	3.24		13.50
3030	1-1/2"	"	0.036	13.00	3.54		16.50
3040	2"	"	0.040	17.00	3.89		21.00
3070	3"	"	0.044	29.00	4.32		33.25
3080	4"	"	0.050	50.00	4.87		55.00

FACILITY WATER DISTRIBUTION

ID Code	Description		Output		Unit Costs			
	Component Descriptions	Unit of Meas.	Manhr / Unit	Material Cost	Labor Cost	Equipment Cost	Total Cost
22 - 11161	**COPPER PIPE, Cont'd...**						**22 - 11161**
3090	6"	LF	0.057	200	5.56		210
4000	Refrigeration tubing, copper, sealed						
4010	1/8"	LF	0.032	0.93	3.11		4.04
4020	3/16"	"	0.033	1.08	3.24		4.32
4030	1/4"	"	0.035	1.30	3.38		4.68
6000	Type L copper						
6090	1/4"	LF	0.024	1.67	2.29		3.96
6095	3/8"	"	0.024	2.56	2.29		4.85
6100	1/2"	"	0.025	2.98	2.43		5.41
6190	3/4"	"	0.027	4.76	2.59		7.35
6240	1"	"	0.029	7.16	2.78		9.94
6580	Type M copper						
6595	3/8"	LF	0.024	1.81	2.29		4.10
6600	1/2"	"	0.025	2.10	2.43		4.53
6620	3/4"	"	0.027	3.43	2.59		6.02
6630	1"	"	0.029	5.57	2.78		8.35
6655	1-1/2"	"	0.033	11.00	3.24		14.25
6685	3-1/2"	"	0.043	46.50	4.21		51.00
7000	Type K tube, coil, material only						
7020	1/4" x 60'	EA					150
7030	1/2" x 60'	"					310
7040	1/2" x 100'	"					510
7050	3/4" x 60'	"					570
7060	3/4" x 100'	"					950
7070	1" x 60'	"					740
7080	1" x 100'	"					1,240
7150	Type L tube, coil						
7170	1/4" x 60'	EA					120
7180	3/8" x 60'	"					190
7190	1/2" x 60'	"					250
7200	1/2" x 100'	"					420
7210	3/4" x 60'	"					400
7220	3/4" x 100'	"					670
7230	1" x 60'	"					580
7240	1" x 100'	"					970

FACILITY WATER DISTRIBUTION

	Description		Output		Unit Costs			
ID Code	Component Descriptions		Unit of Meas.	Manhr / Unit	Material Cost	Labor Cost	Equipment Cost	Total Cost

22 - 11162 — COPPER FITTINGS — 22 - 11162

ID Code	Component Descriptions	Unit of Meas.	Manhr / Unit	Material Cost	Labor Cost	Equipment Cost	Total Cost
0460	Coupling, with stop						
0470	1/4"	EA	0.267	0.95	26.00		27.00
0480	3/8"	"	0.320	1.24	31.25		32.50
0485	1/2"	"	0.348	0.99	33.75		34.75
0490	5/8"	"	0.400	2.87	39.00		41.75
0495	3/4"	"	0.444	1.97	43.25		45.25
0498	1"	"	0.471	4.06	45.75		49.75
0520	Reducing coupling						
0530	1/4" x 1/8"	EA	0.320	2.54	31.25		33.75
0540	3/8" x 1/4"	"	0.348	2.79	33.75		36.50
0545	1/2" x						
0550	3/8"	EA	0.400	2.10	39.00		41.00
0560	1/4"	"	0.400	2.54	39.00		41.50
0570	1/8"	"	0.400	2.80	39.00		41.75
0575	3/4" x						
0580	3/8"	EA	0.444	4.50	43.25		47.75
0590	1/2"	"	0.444	3.56	43.25		46.75
0595	1" x						
0600	3/8"	EA	0.500	8.08	48.75		57.00
0610	1" x 1/2"	"	0.500	7.82	48.75		57.00
0620	1" x 3/4"	"	0.500	6.59	48.75		55.00
0850	Slip coupling						
0860	1/4"	EA	0.267	0.78	26.00		26.75
0870	1/2"	"	0.320	1.31	31.25		32.50
0880	3/4"	"	0.400	2.74	39.00		41.75
0890	1"	"	0.444	5.82	43.25		49.00
1060	Coupling with drain						
1070	1/2"	EA	0.400	10.00	39.00		49.00
1080	3/4"	"	0.444	14.75	43.25		58.00
1090	1"	"	0.500	18.25	48.75		67.00
1110	Reducer						
1120	3/8" x 1/4"	EA	0.320	2.85	31.25		34.00
1130	1/2" x 3/8"	"	0.320	2.29	31.25		33.50
1135	3/4" x						
1140	1/4"	EA	0.364	4.65	35.50		40.25
1150	3/8"	"	0.364	4.86	35.50		40.25
1160	1/2"	"	0.364	5.06	35.50		40.50
1165	1" x						

FACILITY WATER DISTRIBUTION

ID Code	Component Descriptions	Unit of Meas.	Manhr / Unit	Material Cost	Labor Cost	Equipment Cost	Total Cost
	Description	**Output**		**Unit Costs**			

22 - 11162   **COPPER FITTINGS, Cont'd...**   **22 - 11162**

ID Code	Component Descriptions	Unit of Meas.	Manhr / Unit	Material Cost	Labor Cost	Equipment Cost	Total Cost
1170	1/2"	EA	0.400	6.99	39.00		46.00
1180	3/4"	"	0.400	5.36	39.00		44.25
1415	Female adapters						
1430	1/4"	EA	0.320	5.92	31.25		37.25
1440	3/8"	"	0.364	6.06	35.50		41.50
1450	1/2"	"	0.400	2.88	39.00		42.00
1460	3/4"	"	0.444	3.95	43.25		47.25
1470	1"	"	0.444	9.16	43.25		52.00
1540	Increasing female adapters						
1545	1/8" x						
1550	3/8"	EA	0.320	7.27	31.25		38.50
1560	1/2"	"	0.320	6.78	31.25		38.00
1570	1/4" x 1/2"	"	0.348	7.10	33.75		40.75
1580	3/8" x 1/2"	"	0.364	7.62	35.50		43.00
1585	1/2" x						
1590	3/4"	EA	0.400	8.08	39.00		47.00
1600	1"	"	0.400	16.25	39.00		55.00
1605	3/4" x						
1610	1"	EA	0.444	17.25	43.25		61.00
1620	1-1/4"	"	0.444	29.25	43.25		73.00
1625	1" x						
1630	1-1/4"	EA	0.444	31.00	43.25		74.00
1640	1-1/2"	"	0.444	34.00	43.25		77.00
1675	Reducing female adapters						
1690	3/8" x 1/4"	EA	0.364	6.55	35.50		42.00
1695	1/2" x						
1700	1/4"	EA	0.400	5.63	39.00		44.75
1710	3/8"	"	0.400	5.63	39.00		44.75
1720	3/4" x 1/2"	"	0.444	7.86	43.25		51.00
1725	1" x						
1730	1/2"	EA	0.444	14.75	43.25		58.00
1740	3/4"	"	0.444	11.75	43.25		55.00
1840	Female fitting adapters						
1850	1/2"	EA	0.400	10.00	39.00		49.00
1860	3/4"	"	0.400	13.00	39.00		52.00
1870	3/4" x 1/2"	"	0.421	15.50	41.00		57.00
1880	1"	"	0.444	17.25	43.25		61.00
1920	Male adapters						

FACILITY WATER DISTRIBUTION

ID Code	Component Descriptions	Unit of Meas.	Manhr / Unit	Material Cost	Labor Cost	Equipment Cost	Total Cost
22 - 11162	**COPPER FITTINGS, Cont'd...**						**22 - 11162**
1930	1/4"	EA	0.364	11.25	35.50		46.75
1940	3/8"	"	0.364	5.63	35.50		41.25
1970	Increasing male adapters						
1980	3/8" x 1/2"	EA	0.364	7.68	35.50		43.25
1985	1/2" x						
1990	3/4"	EA	0.400	6.66	39.00		45.75
2000	1"	"	0.400	15.00	39.00		54.00
2005	3/4" x						
2010	1"	EA	0.421	14.75	41.00		56.00
2020	1-1/4"	"	0.421	18.75	41.00		60.00
2030	1" x 1-1/4"	"	0.444	18.75	43.25		62.00
2110	Reducing male adapters						
2115	1/2" x						
2120	1/4"	EA	0.400	8.30	39.00		47.25
2130	3/8"	"	0.400	6.89	39.00		46.00
2140	3/4" x 1/2"	"	0.421	7.86	41.00		48.75
2145	1" x						
2150	1/2"	EA	0.444	21.50	43.25		65.00
2160	3/4"	"	0.444	17.25	43.25		61.00
2270	Fitting x male adapters						
2280	1/2"	EA	0.400	11.25	39.00		50.00
2290	3/4"	"	0.421	14.50	41.00		56.00
2300	1"	"	0.444	14.75	43.25		58.00
2340	90 ells						
2350	1/8"	EA	0.320	2.12	31.25		33.25
2360	1/4"	"	0.320	3.37	31.25		34.50
2370	3/8"	"	0.364	3.20	35.50		38.75
2372	1/2"	"	0.400	1.06	39.00		40.00
2374	3/4"	"	0.421	2.40	41.00		43.50
2376	1"	"	0.444	5.90	43.25		49.25
2400	Reducing 90 ell						
2410	3/8" x 1/4"	EA	0.364	5.49	35.50		41.00
2415	1/2" x						
2420	1/4"	EA	0.400	7.86	39.00		46.75
2430	3/8"	"	0.400	7.86	39.00		46.75
2440	3/4" x 1/2"	"	0.421	6.90	41.00		48.00
2445	1" x						
2450	1/2"	EA	0.444	12.00	43.25		55.00

FACILITY WATER DISTRIBUTION

ID Code	Component Descriptions	Unit of Meas.	Manhr / Unit	Material Cost	Labor Cost	Equipment Cost	Total Cost
	Description	**Output**		**Unit Costs**			
22 - 11162	**COPPER FITTINGS, Cont'd...**						**22 - 11162**
2500	3/4"	EA	0.444	11.25	43.25		55.00
2655	Street ells, copper						
2670	1/4"	EA	0.320	5.69	31.25		37.00
2680	3/8"	"	0.364	3.92	35.50		39.50
2690	1/2"	"	0.400	1.58	39.00		40.50
2700	3/4"	"	0.421	3.33	41.00		44.25
2710	1"	"	0.444	8.62	43.25		52.00
2780	Female, 90 ell						
2790	1/2"	EA	0.400	1.07	39.00		40.00
2800	3/4"	"	0.421	2.42	41.00		43.50
2810	1"	"	0.444	5.96	43.25		49.25
2850	Female increasing, 90 ell						
2860	3/8" x 1/2"	EA	0.364	12.00	35.50		47.50
2865	1/2" x						
2870	3/4"	EA	0.400	8.33	39.00		47.25
2880	1"	"	0.400	17.00	39.00		56.00
2890	3/4" x 1"	"	0.421	15.25	41.00		56.00
2900	1" x 1-1/4"	"	0.444	39.00	43.25		82.00
2920	Female reducing, 90 ell						
2930	1/2" x 3/8"	EA	0.400	13.25	39.00		52.00
2940	3/4" x 1/2"	"	0.421	14.75	41.00		56.00
2945	1" x						
2950	1/2"	EA	0.444	20.75	43.25		64.00
2960	3/4"	"	0.444	22.25	43.25		66.00
3040	Male, 90 ell						
3050	1/4"	EA	0.320	10.25	31.25		41.50
3060	3/8"	"	0.364	11.00	35.50		46.50
3070	1/2"	"	0.400	5.66	39.00		44.75
3080	3/4"	"	0.421	12.75	41.00		54.00
3090	1"	"	0.444	14.75	43.25		58.00
3140	Male, increasing 90 ell						
3145	1/2" x						
3150	3/4"	EA	0.400	20.25	39.00		59.00
3160	1"	"	0.400	39.75	39.00		79.00
3170	3/4" x 1"	"	0.421	38.00	41.00		79.00
3180	1" x 1-1/4"	"	0.444	35.00	43.25		78.00
3200	Male, reducing 90 ell						
3210	1/2" x 3/8"	EA	0.400	11.50	39.00		51.00

FACILITY WATER DISTRIBUTION

ID Code	Component Descriptions	Unit of Meas.	Manhr / Unit	Material Cost	Labor Cost	Equipment Cost	Total Cost
	Description	**Output**		**Unit Costs**			
22 - 11162	**COPPER FITTINGS, Cont'd...**						**22 - 11162**
3220	3/4" x 1/2"	EA	0.421	20.25	41.00		61.00
3225	1" x						
3230	1/2"	EA	0.444	38.00	43.25		81.00
3240	3/4"	"	0.444	36.25	43.25		80.00
3260	Drop ear ells						
3270	1/2"	EA	0.400	7.30	39.00		46.25
3280	Female drop ear ells						
3290	1/2"	EA	0.400	7.30	39.00		46.25
3300	1/2" x 3/8"	"	0.400	12.75	39.00		52.00
3310	3/4"	"	0.421	21.25	41.00		62.00
3320	Female flanged sink ell						
3330	1/2"	EA	0.400	13.25	39.00		52.00
3340	45 ells						
3350	1/4"	EA	0.320	6.09	31.25		37.25
3360	3/8"	"	0.364	4.94	35.50		40.50
3390	45 street ell						
3400	1/4"	EA	0.320	6.89	31.25		38.25
3410	3/8"	"	0.364	7.45	35.50		43.00
3420	1/2"	"	0.400	2.21	39.00		41.25
3430	3/4"	"	0.421	3.33	41.00		44.25
3440	1"	"	0.444	8.79	43.25		52.00
3500	Tee						
3510	1/8"	EA	0.320	5.17	31.25		36.50
3520	1/4"	"	0.320	5.45	31.25		36.75
3530	3/8"	"	0.364	4.17	35.50		39.75
3560	Caps						
3570	1/4"	EA	0.320	0.94	31.25		32.25
3580	3/8"	"	0.364	1.50	35.50		37.00
3610	Test caps						
3620	1/2"	EA	0.400	0.88	39.00		40.00
3630	3/4"	"	0.421	0.99	41.00		42.00
3640	1"	"	0.444	1.77	43.25		45.00
3690	Flush bushing						
3700	1/4" x 1/8"	EA	0.320	1.79	31.25		33.00
3705	1/2" x						
3710	1/4"	EA	0.400	2.43	39.00		41.50
3720	3/8"	"	0.400	2.15	39.00		41.25
3725	3/4" x						

FACILITY WATER DISTRIBUTION

ID Code	Component Descriptions	Unit of Meas.	Manhr / Unit	Material Cost	Labor Cost	Equipment Cost	Total Cost
	Description	**Output**		**Unit Costs**			
22 - 11162	**COPPER FITTINGS, Cont'd...**						**22 - 11162**
3730	3/8"	EA	0.421	4.51	41.00		45.50
3740	1/2"	"	0.421	3.99	41.00		45.00
3745	1" x						
3750	1/2"	EA	0.444	6.88	43.25		50.00
3760	3/4"	"	0.444	6.11	43.25		49.25
3850	Female flush bushing						
3855	1/2" x						
3860	1/2" x 1/8"	EA	0.400	5.43	39.00		44.50
3870	1/4"	"	0.400	5.71	39.00		44.75
3880	Union						
3890	1/4"	EA	0.320	33.75	31.25		65.00
3900	3/8"	"	0.364	46.50	35.50		82.00
3920	Female						
3930	1/2"	EA	0.400	16.25	39.00		55.00
3940	3/4"	"	0.421	16.25	41.00		57.00
3950	Male						
3960	1/2"	EA	0.400	17.50	39.00		57.00
3970	3/4"	"	0.421	23.25	41.00		64.00
3980	1"	"	0.444	51.00	43.25		94.00
3990	45 degree wye						
4000	1/2"	EA	0.400	21.75	39.00		61.00
4010	3/4"	"	0.421	31.25	41.00		72.00
4020	1"	"	0.444	42.00	43.25		85.00
4030	1" x 3/4" x 3/4"	"	0.444	58.00	43.25		100
4060	Twin ells						
4070	1" x 3/4" x 3/4"	EA	0.444	15.25	43.25		59.00
4080	1" x 1" x 1"	"	0.444	15.25	43.25		59.00
4150	90 union ells, male						
4160	1/2"	EA	0.400	23.50	39.00		63.00
4170	3/4"	"	0.421	39.00	41.00		80.00
4180	1"	"	0.444	58.00	43.25		100
4190	DWV fittings, coupling with stop						
4200	1-1/4"	EA	0.471	5.31	45.75		51.00
4210	1-1/2"	"	0.500	6.61	48.75		55.00
4220	1-1/2" x 1-1/4"	"	0.500	10.75	48.75		60.00
4230	2"	"	0.533	9.15	52.00		61.00
4240	2" x 1-1/4"	"	0.533	12.50	52.00		65.00
4250	2" x 1-1/2"	"	0.533	12.25	52.00		64.00

FACILITY WATER DISTRIBUTION

| ID Code | Description | Output | | Unit Costs | | | |
	Component Descriptions	Unit of Meas.	Manhr / Unit	Material Cost	Labor Cost	Equipment Cost	Total Cost
22 - 11162	**COPPER FITTINGS, Cont'd...**						**22 - 11162**
4260	3"	EA	0.667	17.75	65.00		83.00
4270	3" x 1-1/2"	"	0.667	42.25	65.00		110
4280	3" x 2"	"	0.667	40.50	65.00		110
4290	4"	"	0.800	56.00	78.00		130
4300	Slip coupling						
4310	1-1/2"	EA	0.500	10.25	48.75		59.00
4320	2"	"	0.533	12.25	52.00		64.00
4330	3"	"	0.667	22.25	65.00		87.00
4340	90 ells						
4350	1-1/2"	EA	0.500	18.75	48.75		68.00
4360	1-1/2" x 1-1/4"	"	0.500	52.00	48.75		100
4370	2"	"	0.533	34.25	52.00		86.00
4380	2" x 1-1/2"	"	0.533	69.00	52.00		120
4390	3"	"	0.667	.91.00	65.00		160
4400	4"	"	0.800	300	78.00		380
4410	Street, 90 elbows						
4420	1-1/2"	EA	0.500	16.00	48.75		65.00
4430	2"	"	0.533	35.00	52.00		87.00
4440	3"	"	0.667	89.00	65.00		150
4450	4"	"	0.800	220	78.00		300
4460	Female, 90 elbows						
4470	1-1/2"	EA	0.500	15.75	48.75		65.00
4480	2"	"	0.533	30.50	52.00		83.00
4490	Male, 90 elbows						
4500	1-1/2"	EA	0.500	28.00	48.75		77.00
4510	2"	"	0.533	57.00	52.00		110
4520	90 with side inlet						
4530	3" x 3" x 1"	EA	0.667	130	65.00		200
4550	3" x 3" x 1-1/2"	"	0.667	130	65.00		200
4560	3" x 3" x 2"	"	0.667	130	65.00		200
4570	45 ells						
4580	1-1/4"	EA	0.471	10.50	45.75		56.00
4590	1-1/2"	"	0.500	8.61	48.75		57.00
4600	2"	"	0.533	19.75	52.00		72.00
4610	3"	"	0.667	42.00	65.00		110
4620	4"	"	0.800	190	78.00		270
4630	Street, 45 ell						
4640	1-1/2"	EA	0.500	13.75	48.75		63.00

FACILITY WATER DISTRIBUTION

ID Code	Component Descriptions	Unit of Meas.	Manhr / Unit	Material Cost	Labor Cost	Equipment Cost	Total Cost
22 - 11162	**COPPER FITTINGS, Cont'd...**						**22 - 11162**
4650	2"	EA	0.533	24.75	52.00		77.00
4660	3"	"	0.667	71.00	65.00		140
4670	60 ell						
4680	1-1/2"	EA	0.500	21.50	48.75		70.00
4690	2"	"	0.533	40.00	52.00		92.00
4700	3"	"	0.667	89.00	65.00		150
4710	22-1/2 ell						
4720	1-1/2"	EA	0.500	26.25	48.75		75.00
4730	2"	"	0.533	33.75	52.00		86.00
4740	3"	"	0.667	59.00	65.00		120
4750	11-1/4 ell						
4760	1-1/2"	EA	0.500	29.00	48.75		78.00
4770	2"	"	0.533	41.00	52.00		93.00
4780	3"	"	0.667	82.00	65.00		150
4790	Wye						
4800	1-1/4"	EA	0.471	44.25	45.75		90.00
4810	1-1/2"	"	0.500	48.25	48.75		97.00
4820	2"	"	0.533	63.00	52.00		110
4830	2" x 1-1/2" x 1-1/2"	"	0.533	70.00	52.00		120
4840	2" x 1-1/2" x 2"	"	0.533	77.00	52.00		130
4850	2" x 1-1/2" x 2-1/2"	"	0.533	77.00	52.00		130
4860	3"	"	0.667	150	65.00		210
4870	3" x 3" x 1-1/2"	"	0.667	140	65.00		200
4880	3" x 3" x 2"	"	0.667	140	65.00		200
4890	4"	"	0.800	310	78.00		390
4900	4" x 4" x 2"	"	0.800	220	78.00		300
4910	4" x 4" x 3"	"	0.800	220	78.00		300
4920	Sanitary tee						
4930	1-1/4"	EA	0.471	22.50	45.75		68.00
4940	1-1/2"	"	0.500	27.75	48.75		77.00
4950	2"	"	0.533	32.50	52.00		85.00
4960	2" x 1-1/2" x 1-1/2"	"	0.533	52.00	52.00		100
4970	2" x 1-1/2" x 2"	"	0.533	53.00	52.00		110
4980	2" x 2" x 1-1/2"	"	0.533	30.75	52.00		83.00
4990	3"	"	0.667	120	65.00		190
5000	3" x 3" x 1-1/2"	"	0.667	93.00	65.00		160
5010	3" x 3" x 2"	"	0.667	93.00	65.00		160
5020	4"	"	0.800	310	78.00		390

FACILITY WATER DISTRIBUTION

	Description	Output		Unit Costs			
ID Code	Component Descriptions	Unit of Meas.	Manhr / Unit	Material Cost	Labor Cost	Equipment Cost	Total Cost
22 - 11162		**COPPER FITTINGS, Cont'd...**					**22 - 11162**
5030	4" x 4" x 3"	EA	0.800	250	78.00		330
5040	Female sanitary tee						
5050	1-1/2"	EA	0.500	54.00	48.75		100
5060	Long turn tee						
5070	1-1/2"	EA	0.500	54.00	48.75		100
5080	2"	"	0.533	120	52.00		170
5090	3" x 1-1/2"	"	0.667	150	65.00		210
5100	Double wye						
5110	1-1/2"	EA	0.500	79.00	48.75		130
5120	2"	"	0.533	140	52.00		190
5130	2" x 2" x 1-1/2" x 1-1/2"	"	0.533	110	52.00		160
5140	3"	"	0.667	230	65.00		300
5150	3" x 3" x 1-1/2" x 1-1/2"	"	0.667	230	65.00		300
5160	3" x 3" x 2" x 2"	"	0.667	230	65.00		300
5170	4" x 4" x 1-1/2" x 1-1/2"	"	0.800	240	78.00		320
5180	Double sanitary tee						
5190	1-1/2"	EA	0.500	41.00	48.75		90.00
5200	2"	"	0.533	94.00	52.00		150
5210	2" x 2" x 1-1/2"	"	0.533	84.00	52.00		140
5220	3"	"	0.667	110	65.00		180
5230	3" x 3" x 1-1/2" x 1-1/2"	"	0.667	140	65.00		200
5240	3" x 3" x 2" x 2"	"	0.667	120	65.00		190
5250	4" x 4" x 1-1/2" x 1-1/2"	"	0.800	250	78.00		330
5260	Long						
5270	2" x 1-1/2"	EA	0.533	140	52.00		190
5280	Twin elbow						
5290	1-1/2"	EA	0.500	71.00	48.75		120
5300	2"	"	0.533	110	52.00		160
5310	2" x 1-1/2" x 1-1/2"	"	0.533	99.00	52.00		150
5320	Spigot adapter, manoff						
5330	1-1/2" x 2"	EA	0.500	47.00	48.75		96.00
5340	1-1/2" x 3"	"	0.500	57.00	48.75		110
5350	2"	"	0.533	23.25	52.00		75.00
5360	2" x 3"	"	0.533	55.00	52.00		110
5370	2" x 4"	"	0.533	78.00	52.00		130
5380	3"	"	0.667	80.00	65.00		140
5390	3" x 4"	"	0.667	150	65.00		210
5400	4"	"	0.800	120	78.00		200

FACILITY WATER DISTRIBUTION

ID Code	Description — Component Descriptions	Output Unit of Meas.	Output Manhr / Unit	Unit Costs Material Cost	Unit Costs Labor Cost	Unit Costs Equipment Cost	Unit Costs Total Cost
22 - 11162	**COPPER FITTINGS, Cont'd...**						**22 - 11162**
5410	No-hub adapters						
5420	1-1/2" x 2"	EA	0.500	28.50	48.75		77.00
5430	2"	"	0.533	26.75	52.00		79.00
5440	2" x 3"	"	0.533	62.00	52.00		110
5450	3"	"	0.667	54.00	65.00		120
5460	3" x 4"	"	0.667	110	65.00		180
5470	4"	"	0.800	120	78.00		200
5480	Fitting reducers						
5490	1-1/2" x 1-1/4"	EA	0.500	10.00	48.75		59.00
5500	2" x 1-1/2"	"	0.533	15.75	52.00		68.00
5510	3" x 1-1/2"	"	0.667	44.25	65.00		110
5520	3" x 2"	"	0.667	39.50	65.00		100
5530	Slip joint (Desanco)						
5540	1-1/4"	EA	0.471	17.00	45.75		63.00
5550	1-1/2"	"	0.500	17.50	48.75		66.00
5560	1-1/2" x 1-1/4"	"	0.500	18.25	48.75		67.00
5570	Street x slip joint (Desanco)						
5580	1-1/2"	EA	0.500	21.75	48.75		71.00
5590	1-1/2" x 1-1/4"	"	0.500	23.25	48.75		72.00
5600	Flush bushing						
5610	1-1/2" x 1-1/4"	EA	0.500	12.50	48.75		61.00
5620	2" x 1-1/2"	"	0.533	21.50	52.00		74.00
5630	3" x 1-1/2"	"	0.667	38.50	65.00		100
5640	3" x 2"	"	0.667	38.50	65.00		100
5650	Male hex trap bushing						
5660	1-1/4" x 1-1/2"	EA	0.471	18.25	45.75		64.00
5670	1-1/2"	"	0.500	13.50	48.75		62.00
5680	1-1/2" x 2"	"	0.500	20.25	48.75		69.00
5690	2"	"	0.533	15.75	52.00		68.00
5710	Round trap bushing						
5720	1-1/2"	EA	0.500	15.50	48.75		64.00
5730	2"	"	0.533	16.50	52.00		69.00
5740	Female adapter						
5750	1-1/4"	EA	0.471	16.25	45.75		62.00
5760	1-1/2"	"	0.500	25.50	48.75		74.00
5770	1-1/2" x 2"	"	0.500	65.00	48.75		110
5780	2"	"	0.533	34.50	52.00		87.00
5790	2" x 1-1/2"	"	0.533	56.00	52.00		110

FACILITY WATER DISTRIBUTION

ID Code	Component Descriptions	Unit of Meas.	Manhr / Unit	Material Cost	Labor Cost	Equipment Cost	Total Cost
22 - 11162	**COPPER FITTINGS, Cont'd...**						**22 - 11162**
5800	3"	EA	0.667	140	65.00		200
5810	Fitting x female adapter						
5820	1-1/2"	EA	0.500	34.25	48.75		83.00
5830	2"	"	0.533	45.50	52.00		98.00
5840	Male adapters						
5850	1-1/4"	EA	0.471	14.25	45.75		60.00
5860	1-1/4" x 1-1/2"	"	0.471	33.00	45.75		79.00
5870	1-1/2"	"	0.500	16.25	48.75		65.00
5880	1-1/2" x 2"	"	0.500	62.00	48.75		110
5890	2"	"	0.533	27.50	52.00		80.00
5900	2" x 1-1/2"	"	0.533	64.00	52.00		120
5910	3"	"	0.667	140	65.00		200
5920	Male x slip joint adapters						
5930	1-1/2" x 1-1/4"	EA	0.500	26.00	48.75		75.00
5940	Dandy cleanout						
5950	1-1/2"	EA	0.500	45.75	48.75		95.00
5960	2"	"	0.533	54.00	52.00		110
5970	3"	"	0.667	190	65.00		250
5980	End cleanout, flush pattern						
5990	1-1/2" x 1"	EA	0.500	27.75	48.75		77.00
6000	2" x 1-1/2"	"	0.533	33.50	52.00		86.00
6020	3" x 2-1/2"	"	0.667	71.00	65.00		140
6025	Copper caps						
6030	1-1/2"	EA	0.500	9.47	48.75		58.00
6040	2"	"	0.533	17.50	52.00		70.00
6050	Closet flanges						
6060	3"	EA	0.667	41.50	65.00		110
6070	4"	"	0.800	73.00	78.00		150
6080	Drum traps, with cleanout						
6090	1-1/2" x 3" x 6"	EA	0.500	160	48.75		210
6100	P-trap, swivel, with cleanout						
6110	1-1/2"	EA	0.500	98.00	48.75		150
6120	P-trap, solder union						
6130	1-1/2"	EA	0.500	41.00	48.75		90.00
6140	2"	"	0.533	72.00	52.00		120
6150	With cleanout						
6160	1-1/2"	EA	0.500	45.00	48.75		94.00
6170	2"	"	0.533	80.00	52.00		130

FACILITY WATER DISTRIBUTION

ID Code	Description — Component Descriptions	Output — Unit of Meas.	Output — Manhr / Unit	Unit Costs — Material Cost	Unit Costs — Labor Cost	Unit Costs — Equipment Cost	Unit Costs — Total Cost
22 - 11162	**COPPER FITTINGS, Cont'd...**						**22 - 11162**
6180	2" x 1-1/2"	EA	0.533	80.00	52.00		130
6190	Swivel joint, with cleanout						
6200	1-1/2" x 1-1/4"	EA	0.500	57.00	48.75		110
6210	1-1/2"	"	0.500	74.00	48.75		120
6220	2" x 1-1/2"	"	0.533	90.00	52.00		140
6230	Estabrook TY, with inlets						
6240	3", with 1-1/2" inlet	EA	0.667	140	65.00		200
6250	Fine thread adapters						
6260	1/2"	EA	0.400	3.67	39.00		42.75
6270	1/2" x 1/2" IPS	"	0.400	4.13	39.00		43.25
6280	1/2" x 3/4" IPS	"	0.400	6.76	39.00		45.75
6290	1/2" x male	"	0.400	2.55	39.00		41.50
6300	1/2" x female	"	0.400	5.21	39.00		44.25
8000	Copper pipe fittings						
8010	1/2"						
8020	90 deg ell	EA	0.178	1.62	17.25		18.75
8040	45 deg ell	"	0.178	2.04	17.25		19.25
8060	Tee	"	0.229	2.71	22.25		25.00
8100	Cap	"	0.089	1.10	8.65		9.75
8120	Coupling	"	0.178	1.18	17.25		18.50
8160	Union	"	0.200	8.21	19.50		27.75
8200	3/4"						
8220	90 deg ell	EA	0.200	3.54	19.50		23.00
8240	45 deg ell	"	0.200	4.13	19.50		23.75
8260	Tee	"	0.267	5.92	26.00		32.00
8290	Cap	"	0.094	2.15	9.16		11.25
8300	Coupling	"	0.200	2.40	19.50		22.00
8320	Union	"	0.229	12.00	22.25		34.25
8360	1"						
8380	90 deg ell	EA	0.267	8.21	26.00		34.25
8390	45 deg ell	"	0.267	10.75	26.00		36.75
8400	Tee	"	0.320	13.50	31.25		44.75
8420	Cap	"	0.133	4.00	13.00		17.00
8430	Coupling	"	0.267	5.92	26.00		32.00
8450	Union	"	0.267	15.75	26.00		41.75

FACILITY WATER DISTRIBUTION

ID Code	Description / Component Descriptions	Output Unit of Meas.	Manhr / Unit	Material Cost	Labor Cost	Equipment Cost	Total Cost
22 - 11164	**BRASS I.P.S. FITTINGS**						**22 - 11164**
0005	Fittings, iron pipe size, 45 deg ell						
0010	1/8"	EA	0.320	5.11	31.25		36.25
0020	1/4"	"	0.320	4.64	31.25		36.00
0030	3/8"	"	0.364	5.12	35.50		40.50
0040	1/2"	"	0.400	5.91	39.00		45.00
0050	3/4"	"	0.421	9.29	41.00		50.00
0060	1"	"	0.444	15.50	43.25		59.00
0095	90 deg ell						
0100	1/8"	EA	0.320	5.25	31.25		36.50
0110	1/4"	"	0.320	5.34	31.25		36.50
0120	3/8"	"	0.364	5.34	35.50		40.75
0130	1/2"	"	0.400	5.34	39.00		44.25
0140	3/4"	"	0.421	11.25	41.00		52.00
0150	1"	"	0.444	17.25	43.25		61.00
0195	90 deg ell, reducing						
0200	1/4" x 1/8"	EA	0.320	19.50	31.25		51.00
0210	3/8" x 1/8"	"	0.364	19.50	35.50		55.00
0220	3/8" x 1/4"	"	0.364	21.50	35.50		57.00
0230	1/2" x 1/4"	"	0.400	17.50	39.00		57.00
0240	1/2" x 3/8"	"	0.400	16.00	39.00		55.00
0250	3/4" x 1/2"	"	0.421	21.75	41.00		63.00
0260	1" x 3/8"	"	0.444	110	43.25		150
0270	1" x 1/2"	"	0.444	38.25	43.25		82.00
0280	1" x 3/4"	"	0.444	38.25	43.25		82.00
0365	Street ell, 45 deg						
0370	1/2"	EA	0.400	9.90	39.00		49.00
0380	3/4"	"	0.421	14.00	41.00		55.00
0395	90 deg						
0400	1/8"	EA	0.320	7.97	31.25		39.25
0410	1/4"	"	0.320	7.97	31.25		39.25
0420	3/8"	"	0.364	7.97	35.50		43.50
0430	1/2"	"	0.400	7.97	39.00		47.00
0440	3/4"	"	0.421	14.00	41.00		55.00
0450	1"	"	0.444	23.25	43.25		67.00
0490	Tee, 1/8"	"	0.320	7.28	31.25		38.50
0501	1/4"	"	0.320	7.28	31.25		38.50
0510	3/8"	"	0.364	7.28	35.50		42.75
0520	1/2"	"	0.400	7.28	39.00		46.25

FACILITY WATER DISTRIBUTION

ID Code	Description / Component Descriptions	Output / Unit of Meas.	Manhr / Unit	Material Cost	Labor Cost	Equipment Cost	Total Cost
22 - 11164	**BRASS I.P.S. FITTINGS, Cont'd...**						**22 - 11164**
0530	3/4"	EA	0.421	13.25	41.00		54.00
0540	1"	"	0.444	23.50	43.25		67.00
0585	Tee, reducing, 3/8" x						
0590	1/4"	EA	0.364	22.50	35.50		58.00
0600	1/2"	"	0.364	11.00	35.50		46.50
0605	1/2" x						
0610	1/4"	EA	0.400	11.00	39.00		50.00
0620	3/8"	"	0.400	11.00	39.00		50.00
0630	3/4"	"	0.400	15.00	39.00		54.00
0635	3/4" x						
0640	1/4"	EA	0.421	17.25	41.00		58.00
0650	1/2"	"	0.421	17.25	41.00		58.00
0660	1"	"	0.421	31.00	41.00		72.00
0665	1" x						
0670	1/2"	EA	0.444	58.00	43.25		100
0680	3/4"	"	0.444	58.00	43.25		100
0810	Tee, reducing						
0820	1/2" x 3/8" x 1/2"	EA	0.400	11.50	39.00		51.00
0830	3/4" x 1/2" x 1/2"	"	0.421	11.50	41.00		53.00
0840	3/4" x 1/2" x 3/4"	"	0.421	13.00	41.00		54.00
0850	1" x 1/2" x 1/2"	"	0.444	56.00	43.25		99.00
0860	1" x 1/2" x 3/4"	"	0.444	56.00	43.25		99.00
0870	1" x 3/4" x 1/2"	"	0.444	53.00	43.25		96.00
0880	1" x 3/4" x 3/4"	"	0.444	21.50	43.25		65.00
0925	Union						
0930	1/8"	EA	0.320	21.00	31.25		52.00
0940	1/4"	"	0.320	21.00	31.25		52.00
0950	3/8"	"	0.364	21.00	35.50		57.00
0960	1/2"	"	0.400	21.00	39.00		60.00
0970	3/4"	"	0.421	30.25	41.00		71.00
0980	1"	"	0.444	40.00	43.25		83.00
1015	Brass face bushing						
1020	3/8" x 1/4"	EA	0.364	12.50	35.50		48.00
1030	1/2" x 3/8"	"	0.400	12.50	39.00		52.00
1040	3/4" x 1/2"	"	0.421	15.75	41.00		57.00
1050	1" x 3/4"	"	0.444	34.25	43.25		78.00
1070	Hex bushing, 1/4" x 1/8"	"	0.320	4.40	31.25		35.75
1075	1/2" x						

FACILITY WATER DISTRIBUTION

ID Code	Component Descriptions	Unit of Meas.	Manhr / Unit	Material Cost	Labor Cost	Equipment Cost	Total Cost
22 - 11164	**BRASS I.P.S. FITTINGS, Cont'd...**						**22 - 11164**
1090	1/4"	EA	0.400	4.40	39.00		43.50
1100	3/8"	"	0.400	4.40	39.00		43.50
1105	5/8" x						
1110	1/8"	EA	0.400	8.14	39.00		47.25
1120	1/4"	"	0.400	8.14	39.00		47.25
1125	3/4" x						
1130	1/8"	EA	0.421	7.22	41.00		48.25
1140	1/4"	"	0.421	7.22	41.00		48.25
1150	3/8"	"	0.421	5.91	41.00		47.00
1160	1/2"	"	0.421	5.91	41.00		47.00
1165	1" x						
1170	1/4"	EA	0.444	11.25	43.25		55.00
1180	3/8"	"	0.444	11.25	43.25		55.00
1190	1/2"	"	0.444	9.21	43.25		52.00
1200	3/4"	"	0.444	9.21	43.25		52.00
1415	Caps						
1420	1/8"	EA	0.320	4.73	31.25		36.00
1430	1/4"	"	0.320	5.24	31.25		36.50
1440	3/8"	"	0.364	5.50	35.50		41.00
1450	1/2"	"	0.400	5.02	39.00		44.00
1460	3/4"	"	0.421	7.90	41.00		49.00
1470	1"	"	0.444	14.25	43.25		58.00
1505	Couplings						
1510	1/8"	EA	0.320	7.17	31.25		38.50
1520	1/4"	"	0.320	7.17	31.25		38.50
1530	3/8"	"	0.364	7.17	35.50		42.75
1540	1/2"	"	0.400	6.29	39.00		45.25
1550	3/4"	"	0.421	9.01	41.00		50.00
1560	1"	"	0.444	12.75	43.25		56.00
1620	Couplings, reducing, 1/4" x 1/8"	"	0.320	18.00	31.25		49.25
1625	3/8" x						
1630	1/8"	EA	0.364	23.50	35.50		59.00
1640	1/4"	"	0.364	7.69	35.50		43.25
1645	1/2" x						
1650	1/8"	EA	0.400	25.00	39.00		64.00
1660	1/4"	"	0.400	6.72	39.00		45.75
1670	3/8"	"	0.400	6.72	39.00		45.75
1675	3/4" x						

FACILITY WATER DISTRIBUTION

ID Code	Description / Component Descriptions	Output / Unit of Meas.	Manhr / Unit	Unit Costs / Material Cost	Labor Cost	Equipment Cost	Total Cost
22 - 11164	**BRASS I.P.S. FITTINGS, Cont'd...**						**22 - 11164**
1680	1/4"	EA	0.421	11.25	41.00		52.00
1690	3/8"	"	0.421	10.00	41.00		51.00
1700	1/2"	"	0.421	9.98	41.00		51.00
1705	1" x						
1710	1/2"	EA	0.421	16.50	41.00		58.00
1720	3/4"	"	0.421	16.50	41.00		58.00
1835	Square head plug, solid						
1840	1/8"	EA	0.320	6.08	31.25		37.25
1850	1/4"	"	0.320	6.08	31.25		37.25
1860	3/8"	"	0.364	6.08	35.50		41.50
1870	1/2"	"	0.400	6.30	39.00		45.25
1880	3/4"	"	0.421	7.23	41.00		48.25
1885	Cored						
1890	1/2"	EA	0.400	3.77	39.00		42.75
1900	3/4"	"	0.421	4.32	41.00		45.25
1910	1"	"	0.444	16.00	43.25		59.00
1965	Countersunk						
1970	1/2"	EA	0.400	9.68	39.00		48.75
1980	3/4"	"	0.421	15.50	41.00		57.00
2005	Locknut						
2010	3/4"	EA	0.421	7.11	41.00		48.00
2020	1"	"	0.444	8.87	43.25		52.00
3010	Close standard red nipple, 1/8"	"	0.320	2.31	31.25		33.50
3015	1/8" x						
3020	1-1/2"	EA	0.320	1.43	31.25		32.75
3030	2"	"	0.320	1.56	31.25		32.75
3040	2-1/2"	"	0.320	1.81	31.25		33.00
3050	3"	"	0.320	1.91	31.25		33.25
3060	3-1/2"	"	0.320	2.11	31.25		33.25
3070	4"	"	0.320	2.38	31.25		33.75
3080	4-1/2"	"	0.320	2.42	31.25		33.75
3090	5"	"	0.320	2.81	31.25		34.00
3100	5-1/2"	"	0.320	3.02	31.25		34.25
3110	6"	"	0.320	3.31	31.25		34.50
3120	1/4" x close	"	0.320	3.63	31.25		35.00
3125	1/4" x						
3130	1-1/2"	EA	0.320	6.30	31.25		37.50
3140	2"	"	0.320	6.68	31.25		38.00

FACILITY WATER DISTRIBUTION

ID Code	Description Component Descriptions	Output Unit of Meas.	Output Manhr / Unit	Unit Costs Material Cost	Unit Costs Labor Cost	Unit Costs Equipment Cost	Unit Costs Total Cost
22 - 11164	**BRASS I.P.S. FITTINGS, Cont'd...**						**22 - 11164**
3150	2-1/2"	EA	0.320	6.98	31.25		38.25
3160	3"	"	0.320	7.32	31.25		38.50
3170	3-1/2"	"	0.320	8.17	31.25		39.50
3180	4"	"	0.320	8.50	31.25		39.75
3190	4-1/2"	"	0.320	9.07	31.25		40.25
3200	5"	"	0.320	9.35	31.25		40.50
3210	5-1/2"	"	0.320	10.25	31.25		41.50
3220	6"	"	0.320	10.50	31.25		41.75
3230	3/8" x close	"	0.364	4.26	35.50		39.75
3235	3/8" x						
3240	1-1/2"	EA	0.364	4.98	35.50		40.50
3250	2"	"	0.364	5.48	35.50		41.00
3260	2-1/2"	"	0.364	6.60	35.50		42.00
3270	3"	"	0.364	8.04	35.50		43.50
3280	3-1/2"	"	0.364	8.79	35.50		44.25
3290	4"	"	0.364	11.25	35.50		46.75
3300	4-1/2"	"	0.364	11.50	35.50		47.00
3310	5"	"	0.364	12.25	35.50		47.75
3320	5-1/2"	"	0.364	13.25	35.50		48.75
3330	6"	"	0.364	14.75	35.50		50.00
3340	1/2" x close	"	0.400	5.61	39.00		44.50
3345	1/2" x						
3350	1-1/2"	EA	0.400	6.99	39.00		46.00
3360	2"	"	0.400	8.51	39.00		47.50
3370	2-1/2"	"	0.400	12.50	39.00		52.00
3380	3"	"	0.400	11.00	39.00		50.00
3390	3-1/2"	"	0.400	12.25	39.00		51.00
3400	4"	"	0.400	12.75	39.00		52.00
3410	4-1/2"	"	0.400	13.75	39.00		53.00
3420	5"	"	0.400	14.25	39.00		53.00
3430	5-1/2"	"	0.400	17.25	39.00		56.00
3440	6"	"	0.400	16.00	39.00		55.00
3450	7-1/2"	"	0.400	48.50	39.00		88.00
3460	8"	"	0.400	48.50	39.00		88.00
3470	3/4" x close	"	0.421	17.25	41.00		58.00
3475	3/4" x						
3480	1-1/2"	EA	0.421	10.50	41.00		52.00
3490	2"	"	0.421	11.00	41.00		52.00

FACILITY WATER DISTRIBUTION

ID Code	Description		Output		Unit Costs			
	Component Descriptions	Unit of Meas.	Manhr / Unit	Material Cost	Labor Cost	Equipment Cost	Total Cost	
22 - 11164	**BRASS I.P.S. FITTINGS, Cont'd...**						**22 - 11164**	
3500	2-1/2"	EA	0.421	12.25	41.00		53.00	
3510	3"	"	0.421	13.25	41.00		54.00	
3520	3-1/2"	"	0.421	14.50	41.00		56.00	
3530	4"	"	0.421	15.50	41.00		57.00	
3540	4-1/2"	"	0.421	16.50	41.00		58.00	
3550	5"	"	0.421	17.25	41.00		58.00	
3560	5-1/2"	"	0.421	19.50	41.00		61.00	
3570	6"	"	0.421	20.25	41.00		61.00	
3580	1" x close	"	0.444	17.50	43.25		61.00	
3585	1" x							
3590	2"	EA	0.444	20.00	43.25		63.00	
3600	2-1/2"	"	0.444	20.25	43.25		64.00	
3610	3"	"	0.444	21.00	43.25		64.00	
3620	3-1/2"	"	0.444	20.00	43.25		63.00	
3630	4"	"	0.444	21.50	43.25		65.00	
3640	4-1/2"	"	0.444	30.00	43.25		73.00	
3650	5"	"	0.444	30.25	43.25		74.00	
3660	5-1/2"	"	0.444	30.25	43.25		74.00	
3670	6"	"	0.444	32.00	43.25		75.00	
22 - 11165	**BRASS FITTINGS**						**22 - 11165**	
1000	Compression fittings, union							
1020	3/8"	EA	0.133	2.85	13.00		15.75	
1030	1/2"	"	0.133	6.16	13.00		19.25	
1040	5/8"	"	0.133	7.84	13.00		20.75	
1050	Union elbow							
1060	3/8"	EA	0.133	9.14	13.00		22.25	
1070	1/2"	"	0.133	15.25	13.00		28.25	
1080	5/8"	"	0.133	20.50	13.00		33.50	
1090	Union tee							
1100	3/8"	EA	0.133	8.57	13.00		21.50	
1120	1/2"	"	0.133	12.50	13.00		25.50	
1130	5/8"	"	0.133	17.25	13.00		30.25	
1140	Male connector							
1150	3/8"	EA	0.133	5.76	13.00		18.75	
1160	1/2"	"	0.133	6.82	13.00		19.75	
1170	5/8"	"	0.133	3.56	13.00		16.50	
1180	Female connector							

FACILITY WATER DISTRIBUTION

ID Code	Component Descriptions	Unit of Meas.	Manhr / Unit	Material Cost	Labor Cost	Equipment Cost	Total Cost
22 - 11165	**BRASS FITTINGS, Cont'd...**						**22 - 11165**
1190	3/8"	EA	0.133	4.86	13.00		17.75
1195	1/2"	"	0.133	6.58	13.00		19.50
1200	5/8"	"	0.133	7.18	13.00		20.25
2000	Brass flare fittings, union						
2020	3/8"	EA	0.129	3.70	12.50		16.25
2030	1/2"	"	0.129	5.11	12.50		17.50
2040	5/8"	"	0.129	6.58	12.50		19.00
2050	90 deg elbow union						
2060	3/8"	EA	0.129	7.95	12.50		20.50
2070	1/2"	"	0.129	11.75	12.50		24.25
2080	5/8"	"	0.129	20.50	12.50		33.00
2090	Three way tee						
2100	3/8"	EA	0.216	7.94	21.00		29.00
2200	1/2"	"	0.216	8.21	21.00		29.25
2210	5/8"	"	0.216	11.00	21.00		32.00
2220	Cross						
2230	3/8"	EA	0.286	12.25	27.75		40.00
2240	1/2"	"	0.286	24.75	27.75		53.00
2250	5/8"	"	0.286	30.00	27.75		58.00
2260	Male connector, half union						
2270	3/8"	EA	0.129	2.83	12.50		15.25
2280	1/2"	"	0.129	4.95	12.50		17.50
2290	5/8"	"	0.129	7.66	12.50		20.25
2300	Female connector, half union						
2310	3/8"	EA	0.129	3.88	12.50		16.50
2320	1/2"	"	0.129	3.60	12.50		16.00
2330	5/8"	"	0.129	6.91	12.50		19.50
2340	Long forged nut						
2350	3/8"	EA	0.129	1.85	12.50		14.25
2360	1/2"	"	0.129	2.69	12.50		15.25
2370	5/8"	"	0.129	9.30	12.50		21.75
2380	Short forged nut						
2390	3/8"	EA	0.129	1.50	12.50		14.00
2400	1/2"	"	0.129	2.04	12.50		14.50
2410	5/8"	"	0.129	2.49	12.50		15.00
3025	Nut, material only						
3030	1/8"	EA					0.34
3040	1/4"	"					0.34

FACILITY WATER DISTRIBUTION

ID Code	Description		Output		Unit Costs			
	Component Descriptions		Unit of Meas.	Manhr / Unit	Material Cost	Labor Cost	Equipment Cost	Total Cost
22 - 11165		**BRASS FITTINGS, Cont'd...**						**22 - 11165**
3050	5/16"		EA					0.42
3060	3/8"		"					0.50
3070	1/2"		"					0.73
3080	5/8"		"					1.57
3105	Sleeve							
3110	1/8"		EA	0.160	0.26	15.50		15.75
3120	1/4"		"	0.160	0.41	15.50		16.00
3130	5/16"		"	0.160	0.24	15.50		15.75
3140	3/8"		"	0.160	0.35	15.50		15.75
3150	1/2"		"	0.160	0.42	15.50		16.00
3160	5/8"		"	0.160	0.61	15.50		16.00
3185	Tee							
3190	1/4"		EA	0.229	5.00	22.25		27.25
3200	5/16"		"	0.229	5.95	22.25		28.25
3205	Male tee							
3210	5/16" x 1/8"		EA	0.229	7.95	22.25		30.25
3215	Female union							
3220	1/8" x 1/8"		EA	0.200	2.02	19.50		21.50
3230	1/4" x 3/8"		"	0.200	3.69	19.50		23.25
3240	3/8" x 1/4"		"	0.200	3.40	19.50		23.00
3250	3/8" x 1/2"		"	0.200	3.85	19.50		23.25
3260	5/8" x 1/2"		"	0.229	6.01	22.25		28.25
3265	Male union, 1/4"							
3270	1/4" x 1/4"		EA	0.200	2.00	19.50		21.50
3300	3/8"		"	0.200	2.65	19.50		22.25
3310	1/2"		"	0.200	3.98	19.50		23.50
3315	5/16" x							
3320	1/8"		EA	0.200	2.17	19.50		21.75
3330	1/4"		"	0.200	2.47	19.50		22.00
3340	3/8"		"	0.200	3.59	19.50		23.00
3345	3/8" x							
3350	1/8"		EA	0.200	2.07	19.50		21.50
3360	1/4"		"	0.200	2.41	19.50		22.00
3370	1/2"		"	0.200	3.36	19.50		22.75
3405	5/8" x							
3410	3/8"		EA	0.229	4.80	22.25		27.00
3420	1/2"		"	0.229	4.14	22.25		26.50
3470	Female elbow, 1/4" x 1/4"		"	0.229	4.62	22.25		26.75

FACILITY WATER DISTRIBUTION

ID Code	Component Descriptions	Unit of Meas.	Manhr / Unit	Material Cost	Labor Cost	Equipment Cost	Total Cost
	Description	**Output**		**Unit Costs**			
22 - 11165		**BRASS FITTINGS, Cont'd...**					**22 - 11165**
3475	5/16" x						
3480	1/8"	EA	0.229	4.97	22.25		27.25
3490	1/4"	"	0.229	7.27	22.25		29.50
3495	3/8" x						
3500	3/8"	EA	0.229	4.01	22.25		26.25
3510	1/2"	"	0.229	3.36	22.25		25.50
3520	Male elbow, 1/8" x 1/8"	"	0.229	4.50	22.25		26.75
3530	3/16" x 1/4"	"	0.229	4.11	22.25		26.25
3535	1/4" x						
3540	1/8"	EA	0.229	2.54	22.25		24.75
3550	1/4"	"	0.229	3.03	22.25		25.25
3560	3/8"	"	0.229	2.58	22.25		24.75
3565	5/16" x						
3570	1/8"	EA	0.229	2.66	22.25		25.00
3580	1/4"	"	0.229	2.96	22.25		25.25
3590	3/8"	"	0.229	5.10	22.25		27.25
3595	3/8" x						
3600	1/8"	EA	0.229	2.55	22.25		24.75
3620	1/4"	"	0.229	3.52	22.25		25.75
3630	3/8"	"	0.229	2.66	22.25		25.00
3640	1/2"	"	0.229	3.55	22.25		25.75
3645	1/2" x						
3650	1/4"	EA	0.267	5.59	26.00		31.50
3660	3/8"	"	0.267	5.19	26.00		31.25
3670	1/2"	"	0.267	4.31	26.00		30.25
3675	5/8" x						
3680	3/8"	EA	0.267	5.62	26.00		31.50
3690	1/2"	"	0.267	5.94	26.00		32.00
3700	3/4"	"	0.267	12.25	26.00		38.25
3735	Union						
3740	1/8"	EA	0.229	2.56	22.25		24.75
3750	3/16"	"	0.229	2.86	22.25		25.00
3760	1/4"	"	0.229	2.18	22.25		24.50
3770	5/16"	"	0.229	2.36	22.25		24.50
3780	3/8"	"	0.229	2.80	22.25		25.00
3805	Reducing union						
3810	3/8" x 1/4"	EA	0.267	3.21	26.00		29.25
3820	5/8" x						

FACILITY WATER DISTRIBUTION

ID Code	Description Component Descriptions	Output Unit of Meas.	Manhr / Unit	Unit Costs Material Cost	Labor Cost	Equipment Cost	Total Cost
22 - 11165	**BRASS FITTINGS, Cont'd...**						**22 - 11165**
3830	3/8"	EA	0.267	5.44	26.00		31.50
3840	1/2"	"	0.267	5.57	26.00		31.50
22 - 11167	**CHROME PLATED FITTINGS**						**22 - 11167**
0005	Fittings						
0010	90 ell						
0020	3/8"	EA	0.200	22.00	19.50		41.50
0030	1/2"	"	0.200	28.50	19.50		48.00
0035	45 ell						
0040	3/8"	EA	0.200	28.50	19.50		48.00
0050	1/2"	"	0.200	37.50	19.50		57.00
0055	Tee						
0060	3/8"	EA	0.267	30.75	26.00		57.00
0070	1/2"	"	0.267	36.50	26.00		63.00
0075	Coupling						
0080	3/8"	EA	0.200	21.50	19.50		41.00
0090	1/2"	"	0.200	21.50	19.50		41.00
0095	Union						
0100	3/8"	EA	0.200	35.50	19.50		55.00
0110	1/2"	"	0.200	36.75	19.50		56.00
0115	Tee						
0130	1/2" x 3/8" x 3/8"	EA	0.267	40.50	26.00		67.00
0140	1/2" x 3/8" x 1/2"	"	0.267	41.25	26.00		67.00
22 - 11168	**PVC/CPVC PIPE**						**22 - 11168**
0900	PVC schedule 40						
1000	1/2" pipe	LF	0.033	0.63	3.24		3.87
1020	3/4" pipe	"	0.036	0.86	3.54		4.40
1040	1" pipe	"	0.040	1.10	3.89		4.99
1060	1-1/4" pipe	"	0.044	1.41	4.32		5.73
1080	1-1/2" pipe	"	0.050	2.11	4.87		6.98
1100	2" pipe	"	0.057	2.68	5.56		8.24
1110	2-1/2" pipe	"	0.067	4.33	6.49		10.75
1120	3" pipe	"	0.080	5.51	7.79		13.25
1130	4" pipe	"	0.100	7.87	9.74		17.50
1190	Fittings, 1/2"						
2000	90 deg ell	EA	0.100	0.58	9.74		10.25
2010	45 deg ell	"	0.100	0.79	9.74		10.50
2020	Tee	"	0.114	0.59	11.25		11.75

FACILITY WATER DISTRIBUTION

ID Code	Description — Component Descriptions	Output — Unit of Meas.	Output — Manhr / Unit	Unit Costs — Material Cost	Unit Costs — Labor Cost	Unit Costs — Equipment Cost	Unit Costs — Total Cost
22 - 11168	**PVC/CPVC PIPE, Cont'd...**						**22 - 11168**
2040	Reducing insert	EA	0.133	0.60	13.00		13.50
2050	Threaded	"	0.100	1.42	9.74		11.25
2060	Male adapter	"	0.133	0.56	13.00		13.50
2070	Female adapter	"	0.100	0.60	9.74		10.25
2080	Coupling	"	0.100	0.44	9.74		10.25
2090	Union	"	0.160	4.73	15.50		20.25
2100	Cap	"	0.133	0.55	13.00		13.50
2101	Flange	"	0.160	10.00	15.50		25.50
2105	3/4"						
2110	90 deg elbow	EA	0.133	0.56	13.00		13.50
2120	45 deg elbow	"	0.133	1.34	13.00		14.25
2130	Tee	"	0.160	0.77	15.50		16.25
2141	Reducing insert	"	0.114	0.56	11.25		11.75
2150	Threaded	"	0.133	0.85	13.00		13.75
2155	1"						
2240	90 deg elbow	EA	0.160	0.99	15.50		16.50
2250	45 deg elbow	"	0.160	1.45	15.50		17.00
2260	Tee	"	0.178	1.32	17.25		18.50
2270	Reducing insert	"	0.160	0.99	15.50		16.50
2280	Threaded	"	0.178	1.32	17.25		18.50
2290	Male adapter	"	0.200	0.92	19.50		20.50
2300	Female adapter	"	0.200	0.79	19.50		20.25
2310	Coupling	"	0.200	0.72	19.50		20.25
2320	Union	"	0.267	7.11	26.00		33.00
2330	Cap	"	0.160	0.79	15.50		16.25
2340	Flange	"	0.267	10.00	26.00		36.00
2345	1-1/4"						
2350	90 deg elbow	EA	0.229	1.71	22.25		24.00
2360	45 deg elbow	"	0.229	2.03	22.25		24.25
2370	Tee	"	0.267	1.98	26.00		28.00
2380	Reducing insert	"	0.267	1.18	26.00		27.25
2390	Threaded	"	0.267	1.98	26.00		28.00
2410	Female adapter	"	0.267	1.25	26.00		27.25
2420	Coupling	"	0.267	1.05	26.00		27.00
2430	Union	"	0.320	16.25	31.25		47.50
2440	Cap	"	0.267	1.12	26.00		27.00
2450	Flange	"	0.320	10.25	31.25		41.50
2455	1-1/2"						

FACILITY WATER DISTRIBUTION

ID Code	Description — Component Descriptions	Output — Unit of Meas.	Output — Manhr / Unit	Unit Costs — Material Cost	Labor Cost	Equipment Cost	Total Cost
22 - 11168	**PVC/CPVC PIPE, Cont'd...**						**22 - 11168**
2460	90 deg elbow	EA	0.229	1.91	22.25		24.25
2470	45 deg elbow	"	0.229	2.82	22.25		25.00
2480	Tee	"	0.267	2.64	26.00		28.75
2490	Reducing insert	"	0.267	1.32	26.00		27.25
2500	Threaded	"	0.267	2.37	26.00		28.25
2510	Male adapter	"	0.267	1.58	26.00		27.50
2520	Female adapter	"	0.267	1.58	26.00		27.50
2530	Coupling	"	0.267	1.18	26.00		27.25
2540	Union	"	0.400	22.25	39.00		61.00
2550	Cap	"	0.267	1.25	26.00		27.25
2560	Flange	"	0.400	17.25	39.00		56.00
2565	2"						
2570	90 deg elbow	EA	0.267	3.03	26.00		29.00
2580	45 deg elbow	"	0.267	3.82	26.00		29.75
2590	Tee	"	0.320	4.02	31.25		35.25
2600	Reducing insert	"	0.320	2.56	31.25		33.75
2610	Threaded	"	0.320	3.41	31.25		34.75
2620	Male adapter	"	0.320	2.11	31.25		33.25
2630	Female adapter	"	0.320	2.17	31.25		33.50
2640	Coupling	"	0.320	1.76	31.25		33.00
2650	Union	"	0.500	30.50	48.75		79.00
2660	Cap	"	0.320	1.65	31.25		33.00
2670	Flange	"	0.500	18.25	48.75		67.00
2675	2-1/2"						
2680	90 deg elbow	EA	0.500	9.14	48.75		58.00
2690	45 deg elbow	"	0.500	13.25	48.75		62.00
2700	Tee	"	0.533	11.75	52.00		64.00
2710	Reducing insert	"	0.533	3.68	52.00		56.00
2720	Threaded	"	0.533	5.26	52.00		57.00
2730	Male adapter	"	0.533	6.12	52.00		58.00
2735	Female adapter	"	0.533	5.06	52.00		57.00
2740	Coupling	"	0.533	3.65	52.00		56.00
2750	Union	"	0.667	40.75	65.00		110
2760	Cap	"	0.500	4.88	48.75		54.00
2770	Flange	"	0.667	24.50	65.00		90.00
2775	3"						
2780	90 deg elbow	EA	0.667	9.88	65.00		75.00
2790	45 deg elbow	"	0.667	12.75	65.00		78.00

FACILITY WATER DISTRIBUTION

ID Code	Description / Component Descriptions	Output / Unit of Meas.	Manhr / Unit	Material Cost	Labor Cost	Equipment Cost	Total Cost
22 - 11168	**PVC/CPVC PIPE, Cont'd...**						**22 - 11168**
2795	Tee	EA	0.727	15.75	71.00		87.00
2800	Reducing insert	"	0.667	4.73	65.00		70.00
2810	Threaded	"	0.667	6.12	65.00		71.00
2820	Male adapter	"	0.667	7.51	65.00		73.00
2825	Female adapter	"	0.667	6.05	65.00		71.00
2830	Coupling	"	0.667	5.59	65.00		71.00
2840	Union	"	0.800	42.75	78.00		120
2850	Cap	"	0.667	4.88	65.00		70.00
2860	Flange	"	0.800	22.50	78.00		100
2865	4"						
2870	90 deg elbow	EA	0.800	17.75	78.00		96.00
2880	45 deg elbow	"	0.800	23.00	78.00		100
2890	Tee	"	0.889	26.25	87.00		110
2900	Reducing insert	"	0.800	10.75	78.00		89.00
2910	Threaded	"	0.800	13.75	78.00		92.00
2920	Male adapter	"	0.800	9.54	78.00		88.00
2925	Female adapter	"	0.800	10.25	78.00		88.00
2930	Coupling	"	0.800	8.15	78.00		86.00
2940	Union	"	1.000	52.00	97.00		150
2950	Cap	"	0.800	11.00	78.00		89.00
2960	Flange	"	1.000	30.25	97.00		130
2965	PVC schedule 80 pipe						
3070	1-1/2" pipe	LF	0.050	2.80	4.87		7.67
3071	2" pipe	"	0.057	3.78	5.56		9.34
3100	3" pipe	"	0.080	7.80	7.79		15.50
3110	4" pipe	"	0.100	10.25	9.74		20.00
3500	Fittings, 1-1/2"						
3600	90 deg elbow	EA	0.267	6.69	26.00		32.75
3610	45 deg elbow	"	0.267	14.75	26.00		40.75
3620	Tee	"	0.400	23.00	39.00		62.00
3630	Reducing insert	"	0.267	4.24	26.00		30.25
3640	Threaded	"	0.267	5.06	26.00		31.00
3650	Male adapter	"	0.267	8.06	26.00		34.00
3660	Female adapter	"	0.267	8.71	26.00		34.75
3670	Coupling	"	0.267	9.30	26.00		35.25
3680	Union	"	0.400	16.75	39.00		56.00
3690	Cap	"	0.267	4.74	26.00		30.75
3700	Flange	"	0.400	11.00	39.00		50.00

FACILITY WATER DISTRIBUTION

ID Code	Description / Component Descriptions	Output Unit of Meas.	Output Manhr / Unit	Unit Costs Material Cost	Unit Costs Labor Cost	Unit Costs Equipment Cost	Unit Costs Total Cost
22 - 11168		**PVC/CPVC PIPE, Cont'd...**					**22 - 11168**
3705	2"						
3709	90 deg elbow	EA	0.320	8.10	31.25		39.25
3710	45 deg elbow	"	0.320	19.00	31.25		50.00
3715	Tee	"	0.500	28.75	48.75		78.00
3720	Reducing insert	"	0.320	6.03	31.25		37.25
3725	Threaded	"	0.320	6.09	31.25		37.25
3730	Male adapter	"	0.320	11.00	31.25		42.25
3735	Female adapter	"	0.320	15.25	31.25		46.50
3740	2-1/2"						
3743	90 deg elbow	EA	0.500	19.00	48.75		68.00
3745	45 deg elbow	"	0.500	40.00	48.75		89.00
3750	Tee	"	0.667	31.25	65.00		96.00
3755	Reducing insert	"	0.500	10.50	48.75		59.00
3760	Threaded	"	0.500	13.00	48.75		62.00
3765	Male adapter	"	0.500	13.25	48.75		62.00
3770	Female adapter	"	0.500	24.00	48.75		73.00
3775	Coupling	"	0.500	13.00	48.75		62.00
3778	Union	"	0.667	36.00	65.00		100
3780	Cap	"	0.500	15.25	48.75		64.00
3785	Flange	"	0.667	19.25	65.00		84.00
3787	3"						
3790	90 deg elbow	EA	0.667	17.00	65.00		82.00
3795	45 deg elbow	"	0.667	48.75	65.00		110
3800	Tee	"	0.800	39.25	78.00		120
3805	Reducing insert	"	0.667	16.75	65.00		82.00
3810	Threaded	"	0.667	24.25	65.00		89.00
3815	Male adapter	"	0.667	14.75	65.00		80.00
3820	Female adapter	"	0.667	27.00	65.00		92.00
3825	Coupling	"	0.667	14.75	65.00		80.00
3830	Union	"	0.800	46.00	78.00		120
3835	Cap	"	0.667	19.25	65.00		84.00
3840	Flange	"	0.800	22.00	78.00		100
3843	4"						
3845	90 deg elbow	EA	0.800	43.50	78.00		120
3850	45 deg elbow	"	0.800	88.00	78.00		170
3855	Tee	"	1.000	45.25	97.00		140
3860	Reducing insert	"	0.800	23.00	78.00		100
3865	Threaded	"	0.800	37.25	78.00		120

FACILITY WATER DISTRIBUTION

ID Code	Description — Component Descriptions	Output — Unit of Meas.	Output — Manhr / Unit	Unit Costs — Material Cost	Unit Costs — Labor Cost	Unit Costs — Equipment Cost	Unit Costs — Total Cost
22 - 11168	**PVC/CPVC PIPE, Cont'd...**						**22 - 11168**
3870	Male adapter	EA	0.800	26.00	78.00		100
3875	Coupling	"	0.800	18.50	78.00		97.00
3880	Union	"	1.000	43.50	97.00		140
3885	Cap	"	0.800	23.50	78.00		100
3888	Flange	"	1.000	30.00	97.00		130
4900	CPVC schedule 40						
5000	1/2" pipe	LF	0.033	1.13	3.24		4.37
5020	3/4" pipe	"	0.036	1.52	3.54		5.06
5040	1" pipe	"	0.040	2.21	3.89		6.10
5060	1-1/4" pipe	"	0.044	2.90	4.32		7.22
5080	1-1/2" pipe	"	0.050	3.51	4.87		8.38
5100	2" pipe	"	0.057	4.67	5.56		10.25
5900	Fittings, CPVC, schedule 80						
6000	1/2", 90 deg ell	EA	0.080	3.44	7.79		11.25
6020	Tee	"	0.133	10.50	13.00		23.50
6060	3/4", 90 deg ell	"	0.080	4.48	7.79		12.25
6080	Tee	"	0.133	5.74	13.00		18.75
6100	1", 90 deg ell	"	0.089	7.09	8.65		15.75
6120	Tee	"	0.145	8.30	14.25		22.50
6140	1-1/4", 90 deg ell	"	0.089	13.00	8.65		21.75
6160	Tee	"	0.145	15.50	14.25		29.75
6200	1-1/2", 90 deg ell	"	0.160	14.25	15.50		29.75
6220	Tee	"	0.200	17.50	19.50		37.00
6300	2", 90 deg ell	"	0.160	15.50	15.50		31.00
6320	Tee	"	0.200	19.75	19.50		39.25
22 - 11169	**STEEL PIPE**						**22 - 11169**
1000	Black steel, extra heavy pipe, threaded						
1030	1/2" pipe	LF	0.032	2.81	3.11		5.92
1100	3/4" pipe	"	0.032	3.64	3.11		6.75
4000	Fittings, malleable iron, threaded, 1/2" pipe						
4010	90 deg ell	EA	0.267	3.37	26.00		29.25
4020	45 deg ell	"	0.267	4.56	26.00		30.50
4030	Tee	"	0.400	3.66	39.00		42.75
4040	Reducing tee	"	0.400	8.22	39.00		47.25
4045	Cap	"	0.160	2.85	15.50		18.25
4050	Coupling	"	0.320	3.81	31.25		35.00
4070	Union	"	0.267	16.00	26.00		42.00

FACILITY WATER DISTRIBUTION

ID Code	Description / Component Descriptions	Output / Unit of Meas.	Output / Manhr / Unit	Unit Costs / Material Cost	Unit Costs / Labor Cost	Unit Costs / Equipment Cost	Unit Costs / Total Cost
22 - 11169	**STEEL PIPE, Cont'd...**						**22 - 11169**
4080	Nipple, 4" long	EA	0.267	3.00	26.00		29.00
4085	3/4" pipe						
4090	90 deg ell	EA	0.267	3.95	26.00		30.00
4100	45 deg ell	"	0.400	6.26	39.00		45.25
4120	Tee	"	0.400	5.31	39.00		44.25
4140	Reducing tee	"	0.267	9.18	26.00		35.25
4150	Cap	"	0.160	3.81	15.50		19.25
4160	Coupling	"	0.267	4.50	26.00		30.50
4170	Union	"	0.267	18.00	26.00		44.00
4175	Nipple, 4" long	"	0.267	3.47	26.00		29.50
5865	Cast iron fittings						
5867	1/2" pipe						
5870	90 deg. ell	EA	0.267	5.08	26.00		31.00
5880	45 deg. ell	"	0.267	10.25	26.00		36.25
5885	Tee	"	0.400	6.70	39.00		45.75
5890	Reducing tee	"	0.400	12.75	39.00		52.00
5920	3/4" pipe						
5930	90 deg. ell	EA	0.267	5.43	26.00		31.50
5940	45 deg. ell	"	0.267	6.70	26.00		32.75
5950	Tee	"	0.400	8.39	39.00		47.50
5960	Reducing tee	"	0.400	10.75	39.00		49.75
22 - 11170	**GALVANIZED STEEL PIPE**						**22 - 11170**
1000	Galvanized pipe						
1020	1/2" pipe	LF	0.080	4.97	7.79		12.75
1040	3/4" pipe	"	0.100	6.47	9.74		16.25
1200	90 degree ell, 150 lb malleable iron, galvanized						
1210	1/2"	EA	0.160	2.22	15.50		17.75
1220	3/4"	"	0.200	2.95	19.50		22.50
1400	45 degree ell, 150 lb m.i., galv.						
1410	1/2"	EA	0.160	10.25	15.50		25.75
1420	3/4"	"	0.200	4.82	19.50		24.25
1520	Tees, straight, 150 lb m.i., galv.						
1530	1/2"	EA	0.200	2.95	19.50		22.50
1540	3/4"	"	0.229	4.92	22.25		27.25
1640	Tees, reducing, out, 150 lb m.i., galv.						
1650	1/2"	EA	0.200	5.10	19.50		24.50
1660	3/4"	"	0.229	5.91	22.25		28.25

FACILITY WATER DISTRIBUTION

ID Code	Component Descriptions	Unit of Meas.	Manhr / Unit	Material Cost	Labor Cost	Equipment Cost	Total Cost

22 - 11170　GALVANIZED STEEL PIPE, Cont'd...　22 - 11170

1800	Couplings, straight, 150 lb m.i., galv.						
1810	1/2"	EA	0.160	2.73	15.50		18.25
1820	3/4"	"	0.178	3.27	17.25		20.50
1920	Couplings, reducing, 150 lb m.i., galv						
1930	1/2"	EA	0.160	3.18	15.50		18.75
1940	3/4"	"	0.178	3.55	17.25		20.75
2040	Caps, 150 lb m.i., galv.						
2050	1/2"	EA	0.080	2.27	7.79		10.00
2060	3/4"	"	0.084	3.00	8.20		11.25
2170	Unions, 150 lb m.i., galv.						
2180	1/2"	EA	0.200	12.75	19.50		32.25
2190	3/4"	"	0.229	14.25	22.25		36.50
2260	Nipples, galvanized steel, 4" long						
2270	1/2"	EA	0.100	3.27	9.74		13.00
2280	3/4"	"	0.107	4.36	10.50		14.75
2360	90 degree reducing ell, 150 lb m.i., galv.						
2370	3/4" x 1/2"	EA	0.160	3.55	15.50		19.00
2380	1" x 3/4"	"	0.178	4.82	17.25		22.00
2550	Square head plug (C.I.)						
2560	1/2"	EA	0.089	2.25	8.65		11.00
2570	3/4"	"	0.100	5.00	9.74		14.75

22 - 11191　BACKFLOW PREVENTERS　22 - 11191

0080	Backflow preventer, flanged, cast iron, with valves						
0100	3" pipe	EA	4.000	4,030	390		4,420
0120	4" pipe	"	4.444	4,700	430		5,130
1900	Threaded						
2000	3/4" pipe	EA	0.500	640	48.75		690
2020	2" pipe	"	0.800	1,120	78.00		1,200

22 - 11196　VACUUM BREAKERS　22 - 11196

1000	Vacuum breaker, atmospheric, threaded connection						
1010	3/4"	EA	0.320	55.00	31.25		86.00
1018	Anti-siphon, brass						
1020	3/4"	EA	0.320	60.00	31.25		91.00

FACILITY POTABLE-WATER STORAGE TANKS

ID Code	Description — Component Descriptions	Unit of Meas.	Manhr / Unit	Material Cost	Labor Cost	Equipment Cost	Total Cost
22 - 12004	**STORAGE TANKS**						**22 - 12004**
0980	Hot water storage tank, cement lined						
1000	10 gallon	EA	2.667	540	260		800
1020	70 gallon	"	4.000	1,700	390		2,090

FACILITY SANITARY SEWERAGE

ID Code	Description — Component Descriptions	Unit of Meas.	Manhr / Unit	Material Cost	Labor Cost	Equipment Cost	Total Cost
22 - 13160	**C.I. PIPE, ABOVE GROUND**						**22 - 13160**
0980	No hub pipe						
1000	1-1/2" pipe	LF	0.057	12.25	5.56		17.75
1010	2" pipe	"	0.067	10.75	6.49		17.25
1100	3" pipe	"	0.080	15.00	7.79		22.75
1200	4" pipe	"	0.133	19.50	13.00		32.50
4980	No hub fittings, 1-1/2" pipe						
5000	1/4 bend	EA	0.267	13.50	26.00		39.50
5060	1/8 bend	"	0.267	11.50	26.00		37.50
5100	Sanitary tee	"	0.400	19.00	39.00		58.00
5120	Sanitary cross	"	0.400	25.50	39.00		65.00
5140	Plug	"					7.51
5180	Coupling	"					27.00
5200	Wye	"	0.400	23.50	39.00		63.00
5270	Tapped tee	"	0.267	25.00	26.00		51.00
5300	P-trap	"	0.267	21.25	26.00		47.25
5360	Tapped cross	"	0.267	28.25	26.00		54.00
5370	2" pipe						
5380	1/4 bend	EA	0.320	17.00	31.25		48.25
5440	1/8 bend	"	0.320	13.75	31.25		45.00
5480	Sanitary tee	"	0.533	23.25	52.00		75.00
5500	Sanitary cross	"	0.533	39.25	52.00		91.00
5520	Plug	"					8.14
5560	Coupling	"					25.75
5600	Wye	"	0.667	21.75	65.00		87.00
5640	Double wye	"	0.667	33.75	65.00		99.00
5700	2x1-1/2" wye & 1/8 bend	"	0.500	40.75	48.75		90.00
5740	Double wye & 1/8 bend	"	0.667	33.75	65.00		99.00
5780	Test tee less 2" plug	"	0.320	21.00	31.25		52.00
5795	Tapped tee						
5800	2"x2"	EA	0.320	30.00	31.25		61.00
5820	2"x1-1/2"	"	0.320	28.25	31.25		60.00

FACILITY SANITARY SEWERAGE

ID Code	Component Descriptions	Unit of Meas.	Manhr / Unit	Material Cost	Labor Cost	Equipment Cost	Total Cost
	Description	**Output**		**Unit Costs**			
22 - 13160	**C.I. PIPE, ABOVE GROUND, Cont'd...**						**22 - 13160**
5900	P-trap						
5940	2"x2"	EA	0.320	27.25	31.25		59.00
5950	Tapped cross						
5960	2"x1-1/2"	EA	0.320	38.50	31.25		70.00
5980	3" pipe						
6000	1/4 bend	EA	0.400	23.25	39.00		62.00
6080	1/8 bend	"	0.400	19.50	39.00		59.00
6120	Sanitary tee	"	0.500	28.50	48.75		77.00
6140	3"x2" sanitary tee	"	0.500	25.75	48.75		75.00
6160	3"x1-1/2" sanitary tee	"	0.500	27.00	48.75		76.00
6180	Sanitary cross	"	0.667	61.00	65.00		130
6200	3x2" sanitary cross	"	0.667	54.00	65.00		120
6220	Plug						12.00
6260	Coupling	"					29.50
6280	Wye	"	0.667	31.00	65.00		96.00
6320	3x2" wye	"	0.667	23.25	65.00		88.00
6360	Double wye	"	0.667	62.00	65.00		130
6390	3x2" double wye	"	0.667	53.00	65.00		120
6480	3x2" wye & 1/8 bend	"	0.571	29.25	56.00		85.00
6500	3x1-1/2" wye & 1/8 bend	"	0.571	29.25	56.00		85.00
6540	Double wye & 1/8 bend	"	0.667	62.00	65.00		130
6580	3x2" double wye & 1/8 bend	"	0.667	53.00	65.00		120
6600	3x2" reducer	"	0.364	11.75	35.50		47.25
6640	Test tee, less 3" plug	"	0.400	32.50	39.00		72.00
6660	Plug	"					12.00
6680	3x3" tapped tee	"	0.400	74.00	39.00		110
6690	3x2" tapped tee	"	0.400	40.25	39.00		79.00
6700	3x1-1/2" tapped tee	"	0.400	34.25	39.00		73.00
6720	P-trap	"	0.400	54.00	39.00		93.00
6760	3x2" tapped cross	"	0.400	50.00	39.00		89.00
6780	3x1-1/2" tapped cross	"	0.400	47.50	39.00		87.00
6800	Closet flange, 3-1/2" deep	"	0.200	34.50	19.50		54.00
6810	4" pipe						
6820	1/4 bend	EA	0.400	33.75	39.00		73.00
6900	1/8 bend	"	0.400	24.75	39.00		64.00
6940	Sanitary tee	"	0.667	44.00	65.00		110
6960	4x3" sanitary tee	"	0.667	40.50	65.00		110
6980	4x2" sanitary tee	"	0.667	33.75	65.00		99.00

FACILITY SANITARY SEWERAGE

ID Code	Component Descriptions	Unit of Meas.	Manhr / Unit	Material Cost	Labor Cost	Equipment Cost	Total Cost
22 - 13160	**C.I. PIPE, ABOVE GROUND, Cont'd...**						**22 - 13160**
7000	Sanitary cross	EA	0.800	110	78.00		190
7020	4x3" sanitary cross	"	0.800	93.00	78.00		170
7040	4x2" sanitary cross	"	0.800	77.00	78.00		160
7060	Plug	"					18.75
7100	Coupling	"					28.75
7120	Wye	"	0.667	51.00	65.00		120
7160	4x3" wye	"	0.667	44.00	65.00		110
7190	4x2" wye	"	0.667	32.50	65.00		98.00
7220	Double wye	"	0.800	130	78.00		210
7260	4x3" double wye	"	0.800	79.00	78.00		160
7290	4x2" double wye	"	0.800	70.00	78.00		150
7320	Wye & 1/8 bend	"	0.667	69.00	65.00		130
7360	4x3" wye & 1/8 bend	"	0.667	50.00	65.00		110
7380	4x2" wye & 1/8 bend	"	0.667	39.00	65.00		100
7400	Double wye & 1/8 bend	"	0.800	180	78.00		260
7420	4x3" double wye & 1/8 bend	"	0.800	120	78.00		200
7440	4x2" double wye & 1/8 bend	"	0.800	110	78.00		190
7460	4x3" reducer	"	0.400	18.25	39.00		57.00
7480	4x2" reducer	"	0.400	18.25	39.00		57.00
7500	Test tee, less 4" plug	"	0.400	55.00	39.00		94.00
7510	Plug	"					18.75
7520	4x2" tapped tee	"	0.400	40.25	39.00		79.00
7530	4x1-1/2" tapped tee	"	0.400	35.50	39.00		75.00
7540	P-trap	"	0.400	94.00	39.00		130
7560	4x2" tapped cross	"	0.400	72.00	39.00		110
7570	4x1-1/2" tapped cross	"	0.400	56.00	39.00		95.00
7575	Closet flange						
7580	3" deep	EA	0.400	37.50	39.00		77.00
7590	8" deep	"	0.400	97.00	39.00		140
22 - 13161	**C.I. PIPE, BELOW GROUND**						**22 - 13161**
1010	No hub pipe						
1020	1-1/2" pipe	LF	0.040	13.00	3.89		17.00
1030	2" pipe	"	0.044	13.25	4.32		17.50
1120	3" pipe	"	0.050	18.25	4.87		23.00
1220	4" pipe	"	0.067	23.75	6.49		30.25
5000	Fittings, 1-1/2"						
5010	1/4 bend	EA	0.229	15.50	22.25		37.75

FACILITY SANITARY SEWERAGE

	Description	Output		Unit Costs			
ID Code	Component Descriptions	Unit of Meas.	Manhr / Unit	Material Cost	Labor Cost	Equipment Cost	Total Cost
22 - 13161	**C.I. PIPE, BELOW GROUND, Cont'd...**						**22 - 13161**
5080	1/8 bend	EA	0.229	13.00	22.25		35.25
5160	Plug	"					8.14
5220	Wye	"	0.320	21.75	31.25		53.00
5260	Wye & 1/8 bend	"	0.229	23.25	22.25		45.50
5320	P-trap	"	0.229	25.50	22.25		47.75
5370	2"						
5390	1/4 bend	EA	0.267	17.00	26.00		43.00
5460	1/8 bend	"	0.267	14.50	26.00		40.50
5540	Plug	"					8.14
5660	Double wye	"	0.500	33.75	48.75		83.00
5690	Wye & 1/8 bend	"	0.400	23.50	39.00		63.00
5760	Double wye & 1/8 bend	"	0.500	58.00	48.75		110
5920	P-trap	"	0.267	24.75	26.00		51.00
6000	3"						
6020	1/4 bend	EA	0.320	23.25	31.25		55.00
6100	1/8 bend	"	0.320	19.50	31.25		51.00
6240	Plug	"					12.00
6300	Wye	"	0.500	31.00	48.75		80.00
6340	3x2" wye	"	0.500	23.25	48.75		72.00
6520	Wye & 1/8 bend	"	0.500	37.50	48.75		86.00
6560	Double wye & 1/8 bend	"	0.500	89.00	48.75		140
6590	3x2" double wye & 1/8 bend	"	0.500	68.00	48.75		120
6620	3x2" reducer	"	0.320	11.75	31.25		43.00
6740	P-trap	"	0.320	54.00	31.25		85.00
6820	4"						
6840	1/4 bend	EA	0.320	33.75	31.25		65.00
6920	1/8 bend	"	0.320	24.75	31.25		56.00
7080	Plug	"					18.75
7140	Wye	"	0.500	51.00	48.75		100
7180	4x3" wye	"	0.500	44.00	48.75		93.00
7200	4x2" wye	"	0.500	32.50	48.75		81.00
7240	Double wye	"	0.667	130	65.00		200
7280	4x3" double wye	"	0.667	79.00	65.00		140
7300	4x2" double wye	"	0.667	70.00	65.00		140
7350	Wye & 1/8 bend	"	0.500	69.00	48.75		120
7370	4x3" wye & 1/8 bend	"	0.500	50.00	48.75		99.00
7390	4x2" wye & 1/8 bend	"	0.500	39.00	48.75		88.00
7410	Double wye & 1/8 bend	"	0.667	180	65.00		250

FACILITY SANITARY SEWERAGE

ID Code	Description — Component Descriptions	Output — Unit of Meas.	Output — Manhr / Unit	Unit Costs — Material Cost	Unit Costs — Labor Cost	Unit Costs — Equipment Cost	Unit Costs — Total Cost
22 - 13161	**C.I. PIPE, BELOW GROUND, Cont'd...**						**22 - 13161**
7430	4x3" double wye & 1/8 bend	EA	0.667	120	65.00		190
7450	4x2" double wye & 1/8 bend	"	0.667	110	65.00		180
7470	4x3" reducer	"	0.320	18.25	31.25		49.50
7490	4x2" reducer	"	0.320	18.25	31.25		49.50
22 - 13163	**ABS DWV PIPE**						**22 - 13163**
1480	Schedule 40 ABS						
1500	1-1/2" pipe	LF	0.040	1.81	3.89		5.70
1520	2" pipe	"	0.044	2.42	4.32		6.74
1530	3" pipe	"	0.057	4.97	5.56		10.50
1540	4" pipe	"	0.080	7.04	7.79		14.75
1555	Fittings						
1660	1/8 bend						
1670	1-1/2"	EA	0.160	2.30	15.50		17.75
1680	2"	"	0.200	3.39	19.50		23.00
1690	3"	"	0.267	8.14	26.00		34.25
1700	4"	"	0.320	14.50	31.25		45.75
1730	Tee, sanitary						
1740	1-1/2"	EA	0.267	3.33	26.00		29.25
1750	2"	"	0.320	5.14	31.25		36.50
1760	3"	"	0.400	14.00	39.00		53.00
1770	4"	"	0.500	25.75	48.75		75.00
1800	Tee, sanitary reducing						
1810	2 x 1-1/2 x 1-1/2	EA	0.320	4.73	31.25		36.00
1820	2 x 1-1/2 x 2	"	0.333	4.87	32.50		37.25
1830	2 x 2 x 1-1/2	"	0.364	4.52	35.50		40.00
1840	3 x 3 x 1-1/2	"	0.400	8.20	39.00		47.25
1850	3 x 3 x 2	"	0.444	10.25	43.25		54.00
1860	4 x 4 x 1-1/2	"	0.500	25.50	48.75		74.00
1870	4 x 4 x 2	"	0.571	23.75	56.00		80.00
1880	4 x 4 x 3	"	0.615	20.75	60.00		81.00
1900	Wye						
1910	1-1/2"	EA	0.229	4.87	22.25		27.00
1920	2"	"	0.320	6.81	31.25		38.00
1930	3"	"	0.400	15.50	39.00		55.00
1940	4"	"	0.500	33.50	48.75		82.00
1960	Reducer						
1970	2 x 1-1/2	EA	0.200	3.27	19.50		22.75

FACILITY SANITARY SEWERAGE

ID Code	Component Descriptions	Unit of Meas.	Manhr / Unit	Material Cost	Labor Cost	Equipment Cost	Total Cost
	Description	**Output**		**Unit Costs**			
22 - 13163	**ABS DWV PIPE, Cont'd...**						**22 - 13163**
1980	3 x 1-1/2	EA	0.267	8.42	26.00		34.50
1990	3 x 2	"	0.267	7.17	26.00		33.25
2000	4 x 2	"	0.320	14.50	31.25		45.75
2010	4 x 3	"	0.320	14.75	31.25		46.00
2030	P-trap						
2040	1-1/2"	EA	0.267	7.59	26.00		33.50
2050	2"	"	0.296	10.25	28.75		39.00
2060	3"	"	0.348	39.25	33.75		73.00
2070	4"	"	0.400	80.00	39.00		120
2120	Double sanitary tee						
2130	1-1/2"	EA	0.320	7.37	31.25		38.50
2140	2"	"	0.400	10.75	39.00		49.75
2150	3"	"	0.500	29.25	48.75		78.00
2160	4"	"	0.667	47.00	65.00		110
2200	Long sweep, 1/4 bend						
2210	1-1/2"	EA	0.160	3.83	15.50		19.25
2220	2"	"	0.200	4.87	19.50		24.25
2230	3"	"	0.267	11.75	26.00		37.75
2240	4"	"	0.400	21.75	39.00		61.00
2250	Wye, standard						
2260	1-1/2"	EA	0.267	4.92	26.00		31.00
2270	2"	"	0.320	6.81	31.25		38.00
2280	3"	"	0.400	15.75	39.00		55.00
2290	4"	"	0.500	33.50	48.75		82.00
2300	Wye, reducing						
2310	2 x 1-1/2 x 1-1/2	EA	0.267	9.12	26.00		35.00
2320	2 x 2 x 1-1/2	"	0.320	8.70	31.25		40.00
2340	4 x 4 x 2	"	0.500	18.50	48.75		67.00
2350	4 x 4 x 3	"	0.533	25.50	52.00		78.00
2360	Double wye						
2370	1-1/2"	EA	0.320	11.25	31.25		42.50
2380	2"	"	0.400	13.25	39.00		52.00
2390	3"	"	0.500	34.00	48.75		83.00
2400	4"	"	0.667	69.00	65.00		130
2410	2 x 2 x 1-1/2 x 1-1/2	"	0.400	13.25	39.00		52.00
2420	3 x 3 x 2 x 2	"	0.500	28.00	48.75		77.00
2430	4 x 4 x 3 x 3	"	0.667	65.00	65.00		130
2440	Combination wye and 1/8 bend						

FACILITY SANITARY SEWERAGE

ID Code	Component Descriptions	Unit of Meas.	Manhr / Unit	Material Cost	Labor Cost	Equipment Cost	Total Cost
22 - 13163	**ABS DWV PIPE, Cont'd...**						**22 - 13163**
2450	1-1/2"	EA	0.267	7.79	26.00		33.75
2460	2"	"	0.320	9.39	31.25		40.75
2470	3"	"	0.400	20.25	39.00		59.00
2480	4"	"	0.500	41.25	48.75		90.00
2490	2 x 2 x 1-1/2	"	0.320	15.75	31.25		47.00
2500	3 x 3 x 1-1/2	"	0.400	19.25	39.00		58.00
2510	3 x 3 x 2	"	0.400	13.50	39.00		53.00
2520	4 x 4 x 2	"	0.500	26.75	48.75		76.00
2530	4 x 4 x 3	"	0.500	32.75	48.75		82.00
22 - 13167	**DRAINS, ROOF & FLOOR**						**22 - 13167**
1020	Floor drain, cast iron, with cast iron top						
1030	2"	EA	0.667	220	65.00		290
1040	3"	"	0.667	230	65.00		300
1050	4"	"	0.667	480	65.00		550
1090	Roof drain, cast iron						
1100	2"	EA	0.667	280	65.00		350
1110	3"	"	0.667	290	65.00		350
1120	4"	"	0.667	370	65.00		430
22 - 13168	**TRAPS**						**22 - 13168**
0980	Bucket trap, threaded						
1000	3/4"	EA	0.500	230	48.75		280
1080	Inverted bucket steam trap, threaded						
1100	3/4"	EA	0.500	280	48.75		330
1150	With stainless interior						
1160	1/2"	EA	0.500	170	48.75		220
1180	3/4"	"	0.500	200	48.75		250
1215	Brass interior						
1220	3/4"	EA	0.500	310	48.75		360
1245	Cast steel body, threaded, high temperature						
1250	3/4"	EA	0.500	800	48.75		850
1480	Float trap, 15 psi						
1500	3/4"	EA	0.500	200	48.75		250
1980	Float and thermostatic trap, 15 psi						
2000	3/4"	EA	0.500	210	48.75		260
2135	Steam trap, cast iron body, threaded, 125 psi						
2140	3/4"	EA	0.500	250	48.75		300
2175	Thermostatic trap, low pressure, angle type, 25 psi						

FACILITY SANITARY SEWERAGE

ID Code	Component Descriptions	Unit of Meas.	Manhr / Unit	Material Cost	Labor Cost	Equipment Cost	Total Cost
22 - 13168	**TRAPS, Cont'd...**						**22 - 13168**
2180	1/2"	EA	0.500	77.00	48.75		130
2190	3/4"	"	0.500	130	48.75		180
2235	Cast iron body, threaded, 125 psi						
2240	3/4"	EA	0.500	170	48.75		220
22 - 13192	**CLEANOUTS**						**22 - 13192**
0980	Cleanout, wall						
1000	2"	EA	0.533	240	52.00		290
1020	3"	"	0.533	340	52.00		390
1040	4"	"	0.667	340	65.00		400
1050	Floor						
1060	2"	EA	0.667	220	65.00		290
1080	3"	"	0.667	290	65.00		350
1100	4"	"	0.800	300	78.00		380

PLUMBING EQUIPMENT

ID Code	Component Descriptions	Unit of Meas.	Manhr / Unit	Material Cost	Labor Cost	Equipment Cost	Total Cost
22 - 33001	**DOMESTIC WATER HEATERS**						**22 - 33001**
0900	Water heater, electric						
1000	6 gal	EA	1.333	450	130		580
1020	10 gal	"	1.333	460	130		590
1030	15 gal	"	1.333	450	130		580
1040	20 gal	"	1.600	630	160		790
1050	30 gal	"	1.600	820	160		980
1060	40 gal	"	1.600	890	160		1,050
1070	52 gal	"	2.000	1,200	190		1,390
2980	Oil fired						
3000	20 gal	EA	4.000	1,430	390		1,820
3020	50 gal	"	5.714	2,230	560		2,790
5000	Tankless water heater, natural gas						
5010	Minimum	EA	5.333	770	520		1,290
5020	Average	"	8.000	880	780		1,660
5030	Maximum	"	16.000	990	1,560		2,550
5040	Propane						
5050	Minimum	EA	5.333	660	520		1,180
5060	Average	"	8.000	770	780		1,550
5070	Maximum	"	16.000	880	1,560		2,440
8000	For trim and rough-in						
8100	Minimum	EA	2.667	210	260		470

PLUMBING EQUIPMENT

ID Code	Description — Component Descriptions	Output — Unit of Meas.	Output — Manhr / Unit	Unit Costs — Material Cost	Unit Costs — Labor Cost	Unit Costs — Equipment Cost	Unit Costs — Total Cost
22 - 33001	**DOMESTIC WATER HEATERS, Cont'd...**						**22 - 33001**
8150	Average	EA	4.000	300	390		690
8200	Maximum	"	8.000	860	780		1,640
22 - 33002	**SOLAR WATER HEATERS**						**22 - 33002**
1000	Hydronic system, 100-120 Gallons including material						
1010	Minimum	EA					14,360
1020	Average	"					15,120
1030	Maximum	"					15,880
1040	Direct-Solar, 100-120 Gallons						
1050	Minimum	EA					9,530
1060	Average	"					9,970
1070	Maximum	"					10,420
1080	Indirect-Solar tank, 50-80 Gallons						
1090	Minimum	EA					1,910
1100	Average	"					2,350
1110	Maximum	"					2,800
1120	100-120 Gallons						
1130	Minimum	EA					3,180
1140	Average	"					3,620
1150	Maximum	"					4,070
1160	Solar water collector panel, 3 x 8						
1170	Minimum	EA	1.000	1,020	97.00		1,120
1180	Average	"	1.143	1,050	110		1,160
1190	Maximum	"	1.333	1,080	130		1,210
1200	4 x 7						
1210	Minimum	EA	1.000	1,140	97.00		1,240
1220	Average	"	1.143	1,180	110		1,290
1230	Maximum	"	1.333	1,210	130		1,340
1240	4 x 8						
1250	Minimum	EA	1.000	1,210	97.00		1,310
1260	Average	"	1.143	1,240	110		1,350
1270	Maximum	"	1.333	1,270	130		1,400
1280	4 x 10						
1290	Minimum	EA	1.143	1,400	110		1,510
1300	Average	"	1.333	1,520	130		1,650
1310	Maximum	"	1.600	1,650	160		1,810
1320	Passive tube tank system, 12 Tube						
1330	Minimum	EA	1.000	760	97.00		860

PLUMBING EQUIPMENT

ID Code	Description / Component Descriptions	Output / Unit of Meas.	Manhr / Unit	Unit Costs / Material Cost	Labor Cost	Equipment Cost	Total Cost
22 - 33002	**SOLAR WATER HEATERS, Cont'd...**						**22 - 33002**
1340	Average	EA	1.143	950	110		1,060
1350	Maximum	"	1.333	1,140	130		1,270
1360	24 Tube						
1370	Minimum	EA	1.000	1,140	97.00		1,240
1380	Average	"	1.143	1,330	110		1,440
1390	Maximum	"	1.333	1,520	130		1,650
1400	27 Tube						
1410	Minimum	EA	1.000	1,270	97.00		1,370
1420	Average	"	1.143	1,590	110		1,700
1430	Maximum	"	1.333	1,910	130		2,040

PLUMBING FIXTURES

ID Code	Component Descriptions	Unit of Meas.	Manhr / Unit	Material Cost	Labor Cost	Equipment Cost	Total Cost
22 - 42136	**WATER CLOSETS**						**22 - 42136**
0980	Water closet flush tank, floor mounted						
1000	Minimum	EA	2.000	320	190		510
1010	Average	"	2.667	630	260		890
1020	Maximum	"	4.000	960	390		1,350
1030	Handicapped						
1040	Minimum	EA	2.667	540	260		800
1050	Average	"	4.000	970	390		1,360
1060	Maximum	"	8.000	1,850	780		2,630
8980	For trim and rough-in						
9000	Minimum	EA	2.000	250	190		440
9020	Average	"	2.667	300	260		560
9040	Maximum	"	4.000	400	390		790
22 - 42162	**LAVATORIES**						**22 - 42162**
1980	Lavatory, countertop, porcelain enamel on cast iron						
2000	Minimum	EA	1.600	230	160		390
2010	Average	"	2.000	350	190		540
2020	Maximum	"	2.667	630	260		890
2080	Wall hung, china						
2100	Minimum	EA	1.600	320	160		480
2110	Average	"	2.000	370	190		560
2120	Maximum	"	2.667	930	260		1,190
2280	Handicapped						
2300	Minimum	EA	2.000	520	190		710
2310	Average	"	2.667	600	260		860

PLUMBING FIXTURES

ID Code	Component Descriptions	Unit of Meas.	Manhr / Unit	Material Cost	Labor Cost	Equipment Cost	Total Cost
22 - 42162	**LAVATORIES, Cont'd...**						**22 - 42162**
2320	Maximum	EA	4.000	1,000	390		1,390
8980	For trim and rough-in						
9000	Minimum	EA	2.000	270	190		460
9020	Average	"	2.667	450	260		710
9040	Maximum	"	4.000	560	390		950
22 - 42164	**SINKS**						**22 - 42164**
0980	Service sink, 24"x29"						
1000	Minimum	EA	2.000	770	190		960
1020	Average	"	2.667	960	260		1,220
1040	Maximum	"	4.000	1,410	390		1,800
2000	Kitchen sink, single, stainless steel, single bowl						
2020	Minimum	EA	1.600	340	160		500
2040	Average	"	2.000	390	190		580
2060	Maximum	"	2.667	710	260		970
2070	Double bowl						
2080	Minimum	EA	2.000	390	190		580
2100	Average	"	2.667	430	260		690
2120	Maximum	"	4.000	750	390		1,140
2190	Porcelain enamel, cast iron, single bowl						
2200	Minimum	EA	1.600	260	160		420
2220	Average	"	2.000	350	190		540
2240	Maximum	"	2.667	540	260		800
2250	Double bowl						
2260	Minimum	EA	2.000	370	190		560
2280	Average	"	2.667	510	260		770
2300	Maximum	"	4.000	730	390		1,120
2980	Mop sink, 24"x36"x10"						
3000	Minimum	EA	1.600	590	160		750
3020	Average	"	2.000	710	190		900
3040	Maximum	"	2.667	950	260		1,210
5980	Washing machine box						
6000	Minimum	EA	2.000	53.00	190		240
6040	Average	"	2.667	77.00	260		340
6060	Maximum	"	4.000	93.00	390		480
8980	For trim and rough-in						
9000	Minimum	EA	2.667	350	260		610
9020	Average	"	4.000	530	390		920

PLUMBING FIXTURES

ID Code	Component Descriptions	Unit of Meas.	Manhr / Unit	Material Cost	Labor Cost	Equipment Cost	Total Cost
	Description	**Output**		**Unit Costs**			
22 - 42164	**SINKS, Cont'd...**					**22 - 42164**	
9040	Maximum	EA	5.333	680	520		1,200
22 - 42190	**BATHS**					**22 - 42190**	
0980	Bathtub, 5' long						
1000	Minimum	EA	2.667	640	260		900
1020	Average	"	4.000	1,400	390		1,790
1040	Maximum	"	8.000	3,190	780		3,970
1050	6' long						
1060	Minimum	EA	2.667	720	260		980
1080	Average	"	4.000	1,460	390		1,850
1100	Maximum	"	8.000	4,140	780		4,920
1110	Square tub, whirlpool, 4'x4'						
1120	Minimum	EA	4.000	2,200	390		2,590
1140	Average	"	8.000	3,110	780		3,890
1160	Maximum	"	10.000	9,500	970		10,470
1170	5'x5'						
1180	Minimum	EA	4.000	2,200	390		2,590
1200	Average	"	8.000	3,110	780		3,890
1220	Maximum	"	10.000	9,680	970		10,650
1230	6'x6'						
1240	Minimum	EA	4.000	2,680	390		3,070
1260	Average	"	8.000	3,910	780		4,690
1280	Maximum	"	10.000	11,220	970		12,190
8980	For trim and rough-in						
9000	Minimum	EA	2.667	230	260		490
9020	Average	"	4.000	330	390		720
9040	Maximum	"	8.000	950	780		1,730
22 - 42230	**SHOWERS**					**22 - 42230**	
0980	Shower, fiberglass, 36"x34"x84"						
1000	Minimum	EA	5.714	690	560		1,250
1020	Average	"	8.000	970	780		1,750
1040	Maximum	"	8.000	1,400	780		2,180
2980	Steel, 1 piece, 36"x36"						
3000	Minimum	EA	5.714	640	560		1,200
3020	Average	"	8.000	970	780		1,750
3040	Maximum	"	8.000	1,140	780		1,920
3980	Receptor, molded stone, 36"x36"						
4000	Minimum	EA	2.667	270	260		530

PLUMBING FIXTURES

ID Code	Description / Component Descriptions	Unit of Meas.	Manhr / Unit	Material Cost	Labor Cost	Equipment Cost	Total Cost
22 - 42230	**SHOWERS, Cont'd...**						**22 - 42230**
4020	Average	EA	4.000	450	390		840
4040	Maximum	"	6.667	690	650		1,340
8980	For trim and rough-in						
9000	Minimum	EA	3.636	270	350		620
9020	Average	"	4.444	450	430		880
9040	Maximum	"	8.000	560	780		1,340
22 - 42260	**DISPOSALS & ACCESSORIES**						**22 - 42260**
0040	Disposal, continuous feed						
0050	Minimum	EA	1.600	83.00	160		240
0060	Average	"	2.000	230	190		420
0070	Maximum	"	2.667	440	260		700
0200	Batch feed, 1/2 hp						
0220	Minimum	EA	1.600	320	160		480
0230	Average	"	2.000	640	190		830
0240	Maximum	"	2.667	1,090	260		1,350
1100	Hot water dispenser						
1110	Minimum	EA	1.600	230	160		390
1120	Average	"	2.000	370	190		560
1130	Maximum	"	2.667	580	260		840
1140	Epoxy finish faucet	"	1.600	330	160		490
1160	Lock stop assembly	"	1.000	70.00	97.00		170
1170	Mounting gasket	"	0.667	8.13	65.00		73.00
1180	Tailpipe gasket	"	0.667	1.19	65.00		66.00
1190	Stopper assembly	"	0.800	27.75	78.00		110
1200	Switch assembly, on/off	"	1.333	31.75	130		160
1210	Tailpipe gasket washer	"	0.400	1.27	39.00		40.25
1220	Stop gasket	"	0.444	2.79	43.25		46.00
1230	Tailpipe flange	"	0.400	0.31	39.00		39.25
1240	Tailpipe	"	0.500	3.62	48.75		52.00
22 - 42390	**FAUCETS**						**22 - 42390**
0980	Kitchen						
1000	Minimum	EA	1.333	95.00	130		230
1020	Average	"	1.600	270	160		430
1040	Maximum	"	2.000	330	190		520
1050	Bath						
1060	Minimum	EA	1.333	95.00	130		230
1080	Average	"	1.600	280	160		440

PLUMBING FIXTURES

ID Code	Description Component Descriptions	Output Unit of Meas.	Output Manhr / Unit	Unit Costs Material Cost	Unit Costs Labor Cost	Unit Costs Equipment Cost	Total Cost
22 - 42390	**FAUCETS, Cont'd...**						**22 - 42390**
1100	Maximum	EA	2.000	430	190		620
1110	Lavatory, domestic						
1120	Minimum	EA	1.333	100	130		230
1140	Average	"	1.600	320	160		480
1160	Maximum	"	2.000	530	190		720
1290	Washroom						
1300	Minimum	EA	1.333	130	130		260
1320	Average	"	1.600	320	160		480
1340	Maximum	"	2.000	580	190		770
1350	Handicapped						
1360	Minimum	EA	1.600	140	160		300
1380	Average	"	2.000	420	190		610
1400	Maximum	"	2.667	650	260		910
1410	Shower						
1420	Minimum	EA	1.333	130	130		260
1440	Average	"	1.600	370	160		530
1460	Maximum	"	2.000	2,780	190		2,970
1480	For trim and rough-in						
1500	Minimum	EA	1.600	89.00	160		250
1520	Average	"	2.000	140	190		330
1540	Maximum	"	4.000	230	390		620
22 - 42398	**HOSE BIBBS**						**22 - 42398**
0005	Hose bibb						
0010	1/2"	EA	0.267	10.50	26.00		36.50
0200	3/4"	"	0.267	11.00	26.00		37.00

POOLS AND FOUNTAIN PLUMBING SYSTEMS

ID Code	**22 - 51007** **SOLAR WATER HEATERS, POOLS**						**22 - 51007**
1000	Solar Water Heater, 1000 BTU/SF panel, 4x8	EA	1.000	140	97.00		240
1100	4 x 10	"	1.143	160	110		270
1200	4 x 12	"	1.333	180	130		310
2200	Panel Mounting Kit, 4x8	"	0.400	38.00	39.00		77.00
2300	4 x 10	"	0.444	57.00	43.25		100
2400	4 x 12	"	0.500	70.00	48.75		120

DIVISION 23
HVAC

INSULATION

ID Code	Description		Output		Unit Costs			
	Component Descriptions		Unit of Meas.	Manhr / Unit	Material Cost	Labor Cost	Equipment Cost	Total Cost

23 - 07131   DUCTWORK INSULATION   23 - 07131

ID Code	Component Descriptions	Unit of Meas.	Manhr / Unit	Material Cost	Labor Cost	Equipment Cost	Total Cost
0980	Fiberglass duct insulation, plain blanket						
1000	1-1/2" thick	SF	0.010	0.40	0.97		1.37
1060	2" thick	"	0.013	0.52	1.29		1.81
1500	With vapor barrier						
1520	1-1/2" thick	SF	0.010	0.46	0.97		1.43
1540	2" thick	"	0.013	0.59	1.29		1.88
2000	Rigid with vapor barrier						
2020	2" thick	SF	0.027	2.54	2.59		5.13

HYDRONIC PIPING AND PUMPS

23 - 21137   STRAINERS   23 - 21137

ID Code	Component Descriptions	Unit of Meas.	Manhr / Unit	Material Cost	Labor Cost	Equipment Cost	Total Cost
0980	Strainer, Y pattern, 125 psi, cast iron body, threaded						
1000	3/4"	EA	0.286	13.75	27.75		41.50
1980	250 psi, brass body, threaded						
2000	3/4"	EA	0.320	36.00	31.25		67.00
2130	Cast iron body, threaded						
2140	3/4"	EA	0.320	21.00	31.25		52.00

AIR DISTRIBUTION

23 - 31130   METAL DUCTWORK   23 - 31130

ID Code	Component Descriptions	Unit of Meas.	Manhr / Unit	Material Cost	Labor Cost	Equipment Cost	Total Cost
0090	Rectangular duct						
0100	Galvanized steel						
1000	Minimum	LB	0.073	0.92	7.08		8.00
1010	Average	"	0.089	1.15	8.65		9.80
1020	Maximum	"	0.133	1.76	13.00		14.75
1080	Aluminum						
1100	Minimum	LB	0.160	2.41	15.50		18.00
1120	Average	"	0.200	3.21	19.50		22.75
1140	Maximum	"	0.267	3.98	26.00		30.00
1160	Fittings						
1180	Minimum	EA	0.267	7.62	26.00		33.50
1200	Average	"	0.400	11.50	39.00		51.00
1220	Maximum	"	0.800	16.75	78.00		95.00

AIR DISTRIBUTION

ID Code	Description / Component Descriptions	Output Unit of Meas.	Manhr / Unit	Material Cost	Labor Cost	Equipment Cost	Total Cost
23 - 33130	**DAMPERS**						**23 - 33130**
0980	Horizontal parallel aluminum backdraft damper						
1000	12" x 12"	EA	0.200	58.00	19.50		78.00
1010	16" x 16"	"	0.229	60.00	22.25		82.00
23 - 33460	**FLEXIBLE DUCTWORK**						**23 - 33460**
1010	Flexible duct, 1.25" fiberglass						
1020	5" dia.	LF	0.040	3.47	3.89		7.36
1040	6" dia.	"	0.044	3.86	4.32		8.18
1060	7" dia.	"	0.047	4.77	4.58		9.35
1080	8" dia.	"	0.050	5.00	4.87		9.87
1100	10" dia.	"	0.057	6.66	5.56		12.25
1120	12" dia.	"	0.062	7.27	5.99		13.25
9000	Flexible duct connector, 3" wide fabric	"	0.133	2.42	13.00		15.50
23 - 34001	**EXHAUST FANS**						**23 - 34001**
0160	Belt drive roof exhaust fans						
1020	640 cfm, 2618 fpm	EA	1.000	1,140	97.00		1,240
1030	940 cfm, 2604 fpm	"	1.000	1,480	97.00		1,580
23 - 37131	**DIFFUSERS**						**23 - 37131**
1980	Ceiling diffusers, round, baked enamel finish						
2000	6" dia.	EA	0.267	40.25	26.00		66.00
2020	8" dia.	"	0.333	48.50	32.50		81.00
2040	10" dia.	"	0.333	54.00	32.50		87.00
2060	12" dia.	"	0.333	69.00	32.50		100
2480	Rectangular						
2500	6x6"	EA	0.267	43.00	26.00		69.00
2520	9x9"	"	0.400	52.00	39.00		91.00
2540	12x12"	"	0.400	76.00	39.00		110
2560	15x15"	"	0.400	95.00	39.00		130
2580	18x18"	"	0.400	120	39.00		160
23 - 37134	**REGISTERS AND GRILLES**						**23 - 37134**
0980	Lay in flush mounted, perforated face, return						
1000	6x6/24x24	EA	0.320	54.00	31.25		85.00
1020	8x8/24x24	"	0.320	54.00	31.25		85.00
1040	9x9/24x24	"	0.320	58.00	31.25		89.00
1060	10x10/24x24	"	0.320	63.00	31.25		94.00
1080	12x12/24x24	"	0.320	63.00	31.25		94.00

AIR DISTRIBUTION

ID Code	Description — Component Descriptions	Output — Unit of Meas.	Output — Manhr / Unit	Unit Costs — Material Cost	Unit Costs — Labor Cost	Unit Costs — Equipment Cost	Unit Costs — Total Cost
23 - 37134	**REGISTERS AND GRILLES, Cont'd...**						**23 - 37134**
3040	Rectangular, ceiling return, single deflection						
3060	10x10	EA	0.400	32.25	39.00		71.00
3080	12x12	"	0.400	37.50	39.00		77.00
3100	14x14	"	0.400	45.75	39.00		85.00
3120	16x8	"	0.400	37.50	39.00		77.00
3140	16x16	"	0.400	37.50	39.00		77.00
4980	Wall, return air register						
5000	12x12	EA	0.200	53.00	19.50		73.00
5020	16x16	"	0.200	79.00	19.50		99.00
5040	18x18	"	0.200	93.00	19.50		110
5980	Ceiling, return air grille						
6000	6x6	EA	0.267	31.00	26.00		57.00
6020	8x8	"	0.320	38.50	31.25		70.00
6040	10x10	"	0.320	47.75	31.25		79.00
6980	Ceiling, exhaust grille, aluminum egg crate						
7000	6x6	EA	0.267	21.25	26.00		47.25
7020	8x8	"	0.320	21.25	31.25		53.00
7040	10x10	"	0.320	23.50	31.25		55.00
7060	12x12	"	0.400	29.00	39.00		68.00

CENTRAL HEATING EQUIPMENT

ID Code	Description — Component Descriptions	Output — Unit of Meas.	Output — Manhr / Unit	Unit Costs — Material Cost	Unit Costs — Labor Cost	Unit Costs — Equipment Cost	Unit Costs — Total Cost
23 - 52230	**BOILERS**						**23 - 52230**
0900	Cast iron, gas fired, hot water						
1000	115 mbh	EA	20.000	3,150	1,380	1,310	5,840
1020	175 mbh	"	21.818	3,750	1,510	1,430	6,690
1040	235 mbh	"	24.000	4,800	1,660	1,580	8,030
1130	Steam						
1140	115 mbh	EA	20.000	3,470	1,380	1,310	6,160
1160	175 mbh	"	21.818	4,180	1,510	1,430	7,120
1180	235 mbh	"	24.000	4,950	1,660	1,580	8,180
1980	Electric, hot water						
2000	115 mbh	EA	12.000	5,630	830	790	7,250
2020	175 mbh	"	12.000	6,230	830	790	7,850
2040	235 mbh	"	12.000	7,110	830	790	8,730
2130	Steam						
2140	115 mbh	EA	12.000	7,110	830	790	8,730
2160	175 mbh	"	12.000	8,700	830	790	10,320

CENTRAL HEATING EQUIPMENT

ID Code	Description — Component Descriptions	Output — Unit of Meas.	Output — Manhr / Unit	Unit Costs — Material Cost	Unit Costs — Labor Cost	Unit Costs — Equipment Cost	Unit Costs — Total Cost
23 - 52230	**BOILERS, Cont'd...**						**23 - 52230**
2180	235 mbh	EA	12.000	9,500	830	790	11,120
3980	Oil fired, hot water						
4000	115 mbh	EA	16.000	4,150	1,110	1,050	6,310
4010	175 mbh	"	18.462	5,270	1,280	1,210	7,760
4020	235 mbh	"	21.818	7,280	1,510	1,430	10,220
4190	Steam						
4200	115 mbh	EA	16.000	4,150	1,110	1,050	6,310
4220	175 mbh	"	18.462	5,270	1,280	1,210	7,760
4240	235 mbh	"	21.818	6,720	1,510	1,430	9,660
23 - 54130	**FURNACES**						**23 - 54130**
0980	Electric, hot air						
1000	40 mbh	EA	4.000	940	390		1,330
1020	60 mbh	"	4.211	1,010	410		1,420
1040	80 mbh	"	4.444	1,100	430		1,530
1060	100 mbh	"	4.706	1,240	460		1,700
1080	125 mbh	"	4.848	1,520	470		1,990
1980	Gas fired hot air						
2000	40 mbh	EA	4.000	940	390		1,330
2020	60 mbh	"	4.211	1,000	410		1,410
2040	80 mbh	"	4.444	1,150	430		1,580
2060	100 mbh	"	4.706	1,200	460		1,660
2080	125 mbh	"	4.848	1,320	470		1,790
2980	Oil fired hot air						
3000	40 mbh	EA	4.000	1,250	390		1,640
3020	60 mbh	"	4.211	2,070	410		2,480
3040	80 mbh	"	4.444	2,090	430		2,520
3060	100 mbh	"	4.706	2,120	460		2,580
3080	125 mbh	"	4.848	2,200	470		2,670

CENTRAL COOLING EQUIPMENT

ID Code	Description — Component Descriptions	Output — Unit of Meas.	Output — Manhr / Unit	Material Cost	Labor Cost	Equipment Cost	Total Cost
23 - 63001	**CONDENSING UNITS**						**23 - 63001**
0980	Air cooled condenser, single circuit						
1000	3 ton	EA	1.333	1,990	130		2,120
1030	5 ton	"	1.333	3,000	130		3,130
1480	With low ambient dampers						
1500	3 ton	EA	2.000	2,180	190		2,370
1530	5 ton	"	2.000	3,430	190		3,620

AIR HANDLING

ID Code	Component Descriptions	Unit of Meas.	Manhr / Unit	Material Cost	Labor Cost	Equipment Cost	Total Cost
	Description	**Output**		**Unit Costs**			
23 - 74001	**AIR HANDLING UNITS**						**23 - 74001**
0980	Air handling unit, medium pressure, single zone						
1000	1500 cfm	EA	5.000	4,840	490		5,330
1060	3000 cfm	"	8.889	6,360	870		7,230
8980	Rooftop air handling units						
9000	4950 cfm	EA	8.889	13,910	870		14,780
9060	7370 cfm	"	11.429	17,640	1,110		18,750

HVAC EQUIPMENT

ID Code	Component Descriptions	Unit of Meas.	Manhr / Unit	Material Cost	Labor Cost	Equipment Cost	Total Cost
23 - 81132	**ROOFTOP UNITS**						**23 - 81132**
0980	Packaged, single zone rooftop unit, with roof curb						
1000	2 ton	EA	8.000	4,130	780		4,910
1020	3 ton	"	8.000	4,340	780		5,120
1040	4 ton	"	10.000	4,740	970		5,710

CONVECTION HEATING AND COOLING UNITS

ID Code	Component Descriptions	Unit of Meas.	Manhr / Unit	Material Cost	Labor Cost	Equipment Cost	Total Cost
23 - 82360	**RADIATION UNITS**						**23 - 82360**
1010	Baseboard radiation unit						
1020	1.7 mbh/lf	LF	0.320	90.00	31.25		120
1040	2.1 mbh/lf	"	0.400	120	39.00		160
23 - 82390	**UNIT HEATERS**						**23 - 82390**
0980	Steam unit heater, horizontal						
1000	12,500 btuh, 200 cfm	EA	1.333	620	130		750
1010	17,000 btuh, 300 cfm	"	1.333	820	130		950

RESISTANCE HEATING

ID Code	Component Descriptions	Unit of Meas.	Manhr / Unit	Material Cost	Labor Cost	Equipment Cost	Total Cost
23 - 83330	**ELECTRIC HEATING**						**23 - 83330**
1000	Baseboard heater						
1020	2', 375w	EA	1.000	53.00	90.00		140
1040	3', 500w	"	1.000	62.00	90.00		150
1060	4', 750w	"	1.143	69.00	100		170
1100	5', 935w	"	1.333	98.00	120		220
1120	6', 1125w	"	1.600	120	140		260
1140	7', 1310w	"	1.818	130	160		290
1160	8', 1500w	"	2.000	150	180		330
1180	9', 1680w	"	2.222	160	200		360

RESISTANCE HEATING

ID Code	Description		Output		Unit Costs			
	Component Descriptions	Unit of Meas.	Manhr / Unit	Material Cost	Labor Cost	Equipment Cost	Total Cost
23 - 83330	**ELECTRIC HEATING, Cont'd...**						**23 - 83330**
1200	10', 1875w	EA	2.286	220	210		430
1210	Unit heater, wall mounted						
1215	750w	EA	1.600	210	140		350
1220	1500w	"	1.667	280	150		430
1270	Thermostat						
1280	Integral	EA	0.500	47.25	45.25		93.00
1300	Line voltage	"	0.500	48.50	45.25		94.00
1320	Electric heater connection	"	0.250	2.08	22.50		24.50
2000	Fittings						
2010	Inside corner	EA	0.400	30.50	36.25		67.00
2020	Outside corner	"	0.400	33.25	36.25		70.00
2030	Receptacle section	"	0.400	34.75	36.25		71.00
2040	Blank section	"	0.400	43.00	36.25		79.00
2185	Radiant ceiling heater panels						
2190	500w	EA	1.000	490	90.00		580
2200	750w	"	1.000	540	90.00		630
2340	Unit heater thermostat	"	0.533	76.00	48.25		120
2350	Mounting bracket	"	0.727	78.00	66.00		140
2360	Relay	"	0.615	99.00	56.00		160

DIVISION 26
ELECTRICAL

CONDUCTORS, CONDUIT AND RACEWAYS

ID Code	Description / Component Descriptions	Output		Unit Costs			
		Unit of Meas.	Manhr / Unit	Material Cost	Labor Cost	Equipment Cost	Total Cost
26 - 05134	**COPPER CONDUCTORS**						**26 - 05134**
0980	Copper conductors, type THW, solid						
1000	#14	LF	0.004	0.20	0.36		0.56
1040	#12	"	0.005	0.31	0.45		0.76
1060	#10	"	0.006	0.48	0.54		1.02
2010	THHN-THWN, solid						
2020	#14	LF	0.004	0.20	0.36		0.56
2040	#12	"	0.005	0.31	0.45		0.76
2060	#10	"	0.006	0.48	0.54		1.02
2070	Stranded						
2080	#14	LF	0.004	0.20	0.36		0.56
2100	#12	"	0.005	0.31	0.45		0.76
2120	#10	"	0.006	0.48	0.54		1.02
2140	#8	"	0.008	0.83	0.72		1.55
2160	#6	"	0.009	1.30	0.81		2.11
2180	#4	"	0.010	2.06	0.90		2.96
2200	#2	"	0.012	2.88	1.08		3.96
2220	#1	"	0.014	3.64	1.26		4.90
2240	1/0	"	0.016	4.48	1.44		5.92
2260	2/0	"	0.020	5.54	1.80		7.34
2280	3/0	"	0.025	6.95	2.26		9.21
2300	4/0	"	0.028	8.70	2.53		11.25
6075	Bare stranded wire						
6080	#8	LF	0.008	0.72	0.72		1.44
6090	#6	"	0.010	1.20	0.90		2.10
6100	#4	"	0.010	1.87	0.90		2.77
6110	#2	"	0.011	2.99	0.99		3.98
6120	#1	"	0.014	3.75	1.26		5.01
6215	Type BX solid armored cable						
6220	#14/2	LF	0.025	1.37	2.26		3.63
6230	#14/3	"	0.028	2.17	2.53		4.70
6240	#14/4	"	0.031	3.04	2.78		5.82
6250	#12/2	"	0.028	1.41	2.53		3.94
6260	#12/3	"	0.031	2.26	2.78		5.04
6270	#12/4	"	0.035	3.14	3.14		6.28
6280	#10/2	"	0.031	2.62	2.78		5.40
6290	#10/3	"	0.035	3.75	3.14		6.89
6300	#10/4	"	0.040	5.83	3.61		9.44
6325	Steel type, metal clad cable, solid, with ground						

CONDUCTORS, CONDUIT AND RACEWAYS

ID Code	Description Component Descriptions	Output Unit of Meas.	Manhr / Unit	Unit Costs Material Cost	Labor Cost	Equipment Cost	Total Cost
26 - 05134	**COPPER CONDUCTORS, Cont'd...**						**26 - 05134**
6330	#14/2	LF	0.018	1.13	1.62		2.75
6340	#14/3	"	0.020	1.74	1.80		3.54
6350	#14/4	"	0.023	2.34	2.06		4.40
6360	#12/2	"	0.020	1.17	1.80		2.97
6370	#12/3	"	0.025	1.93	2.26		4.19
6380	#12/4	"	0.030	2.60	2.73		5.33
6390	#10/2	"	0.023	2.41	2.06		4.47
6400	#10/3	"	0.028	3.36	2.53		5.89
6410	#10/4	"	0.033	5.22	3.01		8.23
26 - 05135	**SHEATHED CABLE**						**26 - 05135**
6700	Non-metallic sheathed cable						
6705	Type NM cable with ground						
6710	#14/2	LF	0.015	0.45	1.35		1.80
6720	#12/2	"	0.016	0.63	1.44		2.07
6730	#10/2	"	0.018	0.99	1.60		2.59
6740	#8/2	"	0.020	1.62	1.80		3.42
6750	#6/2	"	0.025	2.56	2.26		4.82
6760	#14/3	"	0.026	0.56	2.33		2.89
6770	#12/3	"	0.027	0.90	2.41		3.31
6780	#10/3	"	0.027	1.42	2.45		3.87
6790	#8/3	"	0.028	2.38	2.49		4.87
6800	#6/3	"	0.028	3.85	2.53		6.38
6810	#4/3	"	0.032	7.98	2.89		10.75
6820	#2/3	"	0.035	12.00	3.14		15.25
6825	Type UF cable with ground						
6830	#14/2	LF	0.016	0.47	1.44		1.91
6840	#12/2	"	0.019	0.71	1.72		2.43
6880	#14/3	"	0.020	0.66	1.80		2.46
6890	#12/3	"	0.022	1.01	1.98		2.99
6925	Type SFU cable, 3 conductor						
6930	#8	LF	0.028	2.03	2.53		4.56
6940	#6	"	0.031	3.55	2.78		6.33
7025	Type SER cable, 4 conductor						
7030	#6	LF	0.036	5.09	3.28		8.37
7040	#4	"	0.039	7.13	3.52		10.75
7115	Flexible cord, type STO cord						
7120	#18/2	LF	0.004	0.86	0.36		1.22

CONDUCTORS, CONDUIT AND RACEWAYS

	Description	Output		Unit Costs			
ID Code	Component Descriptions	Unit of Meas.	Manhr / Unit	Material Cost	Labor Cost	Equipment Cost	Total Cost
26 - 05135	**SHEATHED CABLE, Cont'd...**						**26 - 05135**
7130	#18/3	LF	0.005	1.01	0.45		1.46
7140	#18/4	"	0.006	1.40	0.54		1.94
7150	#16/2	"	0.004	0.99	0.36		1.35
7160	#16/3	"	0.004	0.83	0.40		1.23
7170	#16/4	"	0.005	1.17	0.45		1.62
7180	#14/2	"	0.005	1.56	0.45		2.01
7190	#14/3	"	0.006	1.42	0.55		1.97
7200	#14/4	"	0.007	1.75	0.63		2.38
7210	#12/2	"	0.006	1.97	0.54		2.51
7220	#12/3	"	0.007	1.48	0.60		2.08
7230	#12/4	"	0.008	2.15	0.72		2.87
7240	#10/2	"	0.007	2.45	0.63		3.08
7250	#10/3	"	0.008	2.35	0.72		3.07
7260	#10/4	"	0.009	3.63	0.81		4.44
26 - 05137	**ALUMINUM CONDUCTORS**						**26 - 05137**
0080	Type XHHW, stranded aluminum, 600v						
0100	#8	LF	0.005	0.63	0.45		1.08
1000	#6	"	0.006	0.67	0.54		1.21
1020	#4	"	0.008	0.82	0.72		1.54
1040	#2	"	0.009	1.14	0.81		1.95
1060	1/0	"	0.011	1.79	0.99		2.78
1080	2/0	"	0.012	2.32	1.08		3.40
1081	3/0	"	0.014	2.90	1.26		4.16
1090	4/0	"	0.015	3.23	1.35		4.58
6245	Type SEU cable						
6250	#8/3	LF	0.025	2.67	2.26		4.93
6260	#6/3	"	0.028	2.67	2.53		5.20
6270	#4/3	"	0.035	3.44	3.14		6.58
6280	#2/3	"	0.038	4.58	3.44		8.02
6290	#1/3	"	0.040	6.21	3.61		9.82
6300	1/0-3	"	0.042	6.98	3.80		10.75
6310	2/0-3	"	0.044	8.02	4.02		12.00
6320	3/0-3	"	0.052	11.25	4.66		16.00
6330	4/0-3	"	0.057	11.25	5.16		16.50
6335	Type SER cable with ground						
6340	#8/3	LF	0.028	3.23	2.53		5.76
6350	#6/3	"	0.035	3.65	3.14		6.79

CONDUCTORS, CONDUIT AND RACEWAYS

ID Code	Description / Component Descriptions	Output Unit of Meas.	Manhr / Unit	Material Cost	Labor Cost	Equipment Cost	Total Cost
26 - 05234	**CONTROL CABLE**						**26 - 05234**
0980	Control cable, 600v, #14 THWN, PVC jacket						
1000	2 wire	LF	0.008	0.54	0.72		1.26
1020	4 wire	"	0.010	0.92	0.90		1.82
26 - 05261	**GROUNDING**						**26 - 05261**
0400	Ground rods, copper clad, 1/2" x						
0510	6'	EA	0.667	22.75	60.00		83.00
0520	8'	"	0.727	31.25	66.00		97.00
0535	5/8" x						
0550	6'	EA	0.727	29.75	66.00		96.00
0560	8'	"	1.000	38.50	90.00		130
1060	Ground rod clamp						
1080	5/8"	EA	0.123	11.25	11.25		22.50
2580	Ground rod couplings						
2600	1/2"	EA	0.100	20.50	9.04		29.50
2610	5/8"	"	0.100	28.75	9.04		37.75
2615	Ground rod, driving stud						
2620	1/2"	EA	0.100	16.50	9.04		25.50
2630	5/8"	"	0.100	19.50	9.04		28.50
2645	Ground rod clamps, #8-2 to						
2650	1" pipe	EA	0.200	18.50	18.00		36.50
2660	2" pipe	"	0.250	23.00	22.50		45.50
26 - 05292	**CONDUIT SPECIALTIES**						**26 - 05292**
8005	Rod beam clamp, 1/2"	EA	0.050	9.72	4.52		14.25
8007	Hanger rod						
8010	3/8"	LF	0.040	2.05	3.61		5.66
8020	1/2"	"	0.050	5.11	4.52		9.63
8030	All thread rod						
8040	1/4"	LF	0.030	0.66	2.73		3.39
8060	3/8"	"	0.040	0.75	3.61		4.36
8080	1/2"	"	0.050	1.40	4.52		5.92
8100	5/8"	"	0.080	2.46	7.23		9.69
8120	Hanger channel, 1-1/2"						
8140	No holes	EA	0.030	6.47	2.73		9.20
8160	Holes	"	0.030	8.00	2.73		10.75
8170	Channel strap						
8180	1/2"	EA	0.050	2.01	4.52		6.53
8200	3/4"	"	0.050	2.71	4.52		7.23

CONDUCTORS, CONDUIT AND RACEWAYS

ID Code	Description / Component Descriptions	Output		Unit Costs			
		Unit of Meas.	Manhr / Unit	Material Cost	Labor Cost	Equipment Cost	Total Cost
26 - 05292	**CONDUIT SPECIALTIES, Cont'd...**						**26 - 05292**
8410	Conduit penetrations, roof and wall, 8" thick						
8420	1/2"	EA	0.615		56.00		56.00
8460	3/4"	"	0.615		56.00		56.00
8480	1"	"	0.800		72.00		72.00
9140	Threaded rod couplings						
9150	1/4"	EA	0.050	2.35	4.52		6.87
9160	3/8"	"	0.050	2.49	4.52		7.01
9170	1/2"	"	0.050	2.81	4.52		7.33
9180	5/8"	"	0.050	4.32	4.52		8.84
9190	3/4"	"	0.050	4.73	4.52		9.25
9195	Hex nuts						
9200	Hex nuts, 1/4"	EA	0.050	0.19	4.52		4.71
9210	3/8"	"	0.050	0.29	4.52		4.81
9220	1/2"	"	0.050	0.63	4.52		5.15
9230	5/8"	"	0.050	1.36	4.52		5.88
9240	3/4"	"	0.050	1.80	4.52		6.32
9245	Square nuts						
9250	1/4"	EA	0.050	0.17	4.52		4.69
9260	3/8"	"	0.050	0.32	4.52		4.84
9270	1/2"	"	0.050	0.56	4.52		5.08
9280	5/8"	"	0.050	0.74	4.52		5.26
9290	3/4"	"	0.050	1.30	4.52		5.82
9295	Flat washers, material only						
9300	1/4"	EA					0.19
9310	3/8"	"					0.26
9320	1/2"	"					0.36
9330	5/8"	"					0.74
9340	3/4"	"					1.03
9345	Lockwashers						
9350	1/4"	EA					0.12
9360	3/8"	"					0.20
9370	1/2"	"					0.25
9380	5/8"	"					0.43
9390	3/4"	"					0.74

CONDUCTORS, CONDUIT AND RACEWAYS

ID Code	Description		Output		Unit Costs			
	Component Descriptions	Unit of Meas.	Manhr / Unit	Material Cost	Labor Cost	Equipment Cost	Total Cost
26 - 05334	**SURFACE MOUNTED RACEWAY**						**26 - 05334**
0980	Single Raceway						
1000	3/4" x 17/32" Conduit	LF	0.040	2.78	3.61		6.39
1020	Mounting Strap	EA	0.053	0.75	4.82		5.57
1040	Connector	"	0.053	1.00	4.82		5.82
1060	Elbow						
2000	45 degree	EA	0.050	12.75	4.52		17.25
2020	90 degree	"	0.050	4.05	4.52		8.57
2040	internal	"	0.050	5.09	4.52		9.61
2050	external	"	0.050	4.71	4.52		9.23
2060	Switch	"	0.400	33.00	36.25		69.00
2100	Utility Box	"	0.400	22.25	36.25		59.00
2110	Receptacle	"	0.400	39.00	36.25		75.00
2140	3/4" x 21/32" Conduit	LF	0.040	3.17	3.61		6.78
2160	Mounting Strap	EA	0.053	1.17	4.82		5.99
2180	Connector	"	0.053	1.20	4.82		6.02
2200	Elbow						
2210	45 degree	EA	0.050	9.41	4.52		14.00
2220	90 degree	"	0.050	3.49	4.52		8.01
2240	internal	"	0.050	4.73	4.52		9.25
2260	external	"	0.050	4.73	4.52		9.25
3000	Switch	"	0.400	26.75	36.25		63.00
3010	Utility Box	"	0.400	18.00	36.25		54.00
3020	Receptacle	"	0.400	31.50	36.25		68.00
26 - 05338	**BOXES**						**26 - 05338**
5000	Round cast box, type SEH						
5010	1/2"	EA	0.348	34.50	31.50		66.00
5020	3/4"	"	0.421	34.50	38.00		73.00
5025	SEHC						
5030	1/2"	EA	0.348	41.00	31.50		73.00
5040	3/4"	"	0.421	41.00	38.00		79.00
5045	SEHL						
5050	1/2"	EA	0.348	42.25	31.50		74.00
5060	3/4"	"	0.444	41.00	40.25		81.00
5065	SEHT						
5070	1/2"	EA	0.421	45.25	38.00		83.00
5080	3/4"	"	0.500	45.25	45.25		91.00
5085	SEHX						

CONDUCTORS, CONDUIT AND RACEWAYS

ID Code		Description	Output		Unit Costs			
	Component Descriptions		Unit of Meas.	Manhr / Unit	Material Cost	Labor Cost	Equipment Cost	Total Cost
26 - 05338			**BOXES, Cont'd...**					**26 - 05338**
5090	1/2"		EA	0.500	49.00	45.25		94.00
5100	3/4"		"	0.615	49.00	56.00		110
5110	Blank cover		"	0.145	8.32	13.25		21.50
5120	1/2", hub cover		"	0.145	7.95	13.25		21.25
5130	Cover with gasket		"	0.178	8.70	16.00		24.75
5135	Rectangle, type FS boxes							
5140	1/2"		EA	0.348	17.75	31.50		49.25
5150	3/4"		"	0.400	18.75	36.25		55.00
5160	1"		"	0.500	20.50	45.25		66.00
5165	FSA							
5170	1/2"		EA	0.348	31.75	31.50		63.00
5180	3/4"		"	0.400	29.50	36.25		66.00
5185	FSC							
5190	1/2"		EA	0.348	19.75	31.50		51.00
5200	3/4"		"	0.421	21.50	38.00		60.00
5210	1"		"	0.500	27.00	45.25		72.00
5215	FSL							
5220	1/2"		EA	0.348	31.50	31.50		63.00
5230	3/4"		"	0.400	31.50	36.25		68.00
5235	FSR							
5240	1/2"		EA	0.348	33.00	31.50		65.00
5250	3/4"		"	0.400	33.50	36.25		70.00
5255	FSS							
5260	1/2"		EA	0.348	19.75	31.50		51.00
5270	3/4"		"	0.400	21.50	36.25		58.00
5275	FSLA							
5280	1/2"		EA	0.348	13.50	31.50		45.00
5290	3/4"		"	0.400	15.25	36.25		52.00
5295	FSCA							
5300	1/2"		EA	0.348	39.75	31.50		71.00
5310	3/4"		"	0.400	38.50	36.25		75.00
5315	FSCC							
5320	1/2"		EA	0.400	24.00	36.25		60.00
5330	3/4"		"	0.500	36.00	45.25		81.00
5335	FSCT							
5340	1/2"		EA	0.400	24.00	36.25		60.00
5350	3/4"		"	0.500	30.00	45.25		75.00
5360	1"		"	0.571	24.50	52.00		77.00

CONDUCTORS, CONDUIT AND RACEWAYS

ID Code	Description — Component Descriptions	Output — Unit of Meas.	Output — Manhr / Unit	Unit Costs — Material Cost	Unit Costs — Labor Cost	Unit Costs — Equipment Cost	Total Cost
26 - 05338	**BOXES, Cont'd...**						**26 - 05338**
5365	FST						
5370	1/2"	EA	0.500	35.50	45.25		81.00
5380	3/4"	"	0.571	35.50	52.00		88.00
5385	FSX						
5390	1/2"	EA	0.615	40.50	56.00		97.00
5400	3/4"	"	0.727	37.25	66.00		100
5405	FSCD boxes						
5410	1/2"	EA	0.615	33.75	56.00		90.00
5420	3/4"	"	0.727	35.25	66.00		100
5425	Rectangle, type FS, 2 gang boxes						
5430	1/2"	EA	0.348	37.75	31.50		69.00
5440	3/4"	"	0.400	39.00	36.25		75.00
5450	1"	"	0.500	41.00	45.25		86.00
5455	FSC, 2 gang boxes						
5460	1/2"	EA	0.348	40.00	31.50		72.00
5470	3/4"	"	0.400	44.25	36.25		81.00
5480	1"	"	0.500	54.00	45.25		99.00
5485	FSS, 2 gang boxes						
5490	3/4"	EA	0.400	42.00	36.25		78.00
5495	FS, tandem boxes						
5500	1/2"	EA	0.400	41.75	36.25		78.00
5510	3/4"	"	0.444	42.75	40.25		83.00
5515	FSC, tandem boxes						
5520	1/2"	EA	0.400	56.00	36.25		92.00
5530	3/4"	"	0.444	60.00	40.25		100
5535	FS, three gang boxes						
5540	3/4"	EA	0.444	61.00	40.25		100
5550	1"	"	0.500	68.00	45.25		110
5560	FSS, three gang boxes, 3/4"	"	0.500	79.00	45.25		120
5565	Weatherproof cast aluminum boxes, 1 gang, 3 outlets						
5570	1/2"	EA	0.400	11.25	36.25		47.50
5580	3/4"	"	0.500	12.25	45.25		58.00
5585	2 gang, 3 outlets						
5590	1/2"	EA	0.500	21.50	45.25		67.00
5600	3/4"	"	0.533	23.00	48.25		71.00
5605	1 gang, 4 outlets						
5610	1/2"	EA	0.615	19.50	56.00		76.00
5620	3/4"	"	0.727	21.50	66.00		88.00

CONDUCTORS, CONDUIT AND RACEWAYS

ID Code	Component Descriptions	Unit of Meas.	Manhr / Unit	Material Cost	Labor Cost	Equipment Cost	Total Cost
	Description	**Output**		**Unit Costs**			
26 - 05338		**BOXES, Cont'd...**					**26 - 05338**
5625	2 gang, 4 outlets						
5630	1/2"	EA	0.615	20.50	56.00		77.00
5640	3/4"	"	0.727	22.75	66.00		89.00
5645	1 gang, 5 outlets						
5650	1/2"	EA	0.727	16.25	66.00		82.00
5660	3/4"	"	0.800	19.25	72.00		91.00
5665	2 gang, 5 outlets						
5670	1/2"	EA	0.727	29.25	66.00		95.00
5680	3/4"	"	0.800	35.50	72.00		110
5685	2 gang, 6 outlets						
5690	1/2"	EA	0.851	33.25	77.00		110
5700	3/4"	"	0.899	35.50	81.00		120
5705	2 gang, 7 outlets						
5710	1/2"	EA	1.000	35.25	90.00		130
5720	3/4"	"	1.096	44.00	99.00		140
5730	Weatherproof and type FS box covers, blank, 1 gang	"	0.145	5.12	13.25		18.25
5740	Tumbler switch, 1 gang	"	0.145	10.50	13.25		23.75
5750	1 gang, single recept	"	0.145	6.63	13.25		20.00
5760	Duplex recept	"	0.145	8.45	13.25		21.75
5770	Despard	"	0.145	8.48	13.25		21.75
5780	Red pilot light	"	0.145	40.00	13.25		53.00
5785	SW and						
5790	Single recept	EA	0.200	17.75	18.00		35.75
5800	Duplex recept	"	0.200	14.50	18.00		32.50
5805	2 gang						
5810	Blank	EA	0.182	5.32	16.50		21.75
5820	Tumbler switch	"	0.182	7.00	16.50		23.50
5830	Single recept	"	0.182	7.00	16.50		23.50
5840	Duplex recept	"	0.182	7.00	16.50		23.50
5845	3 gang						
5850	Blank	EA	0.200	12.25	18.00		30.25
5860	Tumbler switch	"	0.200	15.25	18.00		33.25
5865	4 gang						
5870	Tumbler switch	EA	0.250	19.25	22.50		41.75
6325	Box covers						
6330	Surface	EA	0.200	27.00	18.00		45.00
6340	Sealing	"	0.200	29.25	18.00		47.25
6350	Dome	"	0.200	40.50	18.00		59.00

CONDUCTORS, CONDUIT AND RACEWAYS

ID Code	Component Descriptions	Unit of Meas.	Manhr / Unit	Material Cost	Labor Cost	Equipment Cost	Total Cost
26 - 05338	**BOXES, Cont'd...**						**26 - 05338**
6360	1/2" nipple	EA	0.200	52.00	18.00		70.00
6370	3/4" nipple	"	0.200	53.00	18.00		71.00
26 - 05339	**PULL AND JUNCTION BOXES**						**26 - 05339**
1050	4"						
1060	Octagon box	EA	0.114	5.67	10.25		16.00
1070	Box extension	"	0.059	9.55	5.36		15.00
1080	Plaster ring	"	0.059	5.24	5.36		10.50
1100	Cover blank	"	0.059	2.31	5.36		7.67
1120	Square box	"	0.114	8.16	10.25		18.50
1140	Box extension	"	0.059	8.00	5.36		13.25
1160	Plaster ring	"	0.059	4.38	5.36		9.74
1180	Cover blank	"	0.059	2.24	5.36		7.60
1300	Switch and device boxes						
1320	2 gang	EA	0.114	24.75	10.25		35.00
1340	3 gang	"	0.114	43.50	10.25		54.00
1360	4 gang	"	0.160	58.00	14.50		73.00
2000	Device covers						
2020	2 gang	EA	0.059	19.75	5.36		25.00
2040	3 gang	"	0.059	20.25	5.36		25.50
2060	4 gang	"	0.059	27.50	5.36		32.75
2100	Handy box	"	0.114	6.08	10.25		16.25
2120	Extension	"	0.059	5.73	5.36		11.00
2140	Switch cover	"	0.059	3.04	5.36		8.40
2160	Switch box with knockout	"	0.145	9.14	13.25		22.50
2200	Weatherproof cover, spring type	"	0.080	17.00	7.23		24.25
2220	Cover plate, dryer receptacle 1 gang plastic	"	0.100	2.59	9.04		11.75
2240	For 4" receptacle, 2 gang	"	0.100	4.63	9.04		13.75
2260	Duplex receptacle cover plate, plastic	"	0.059	1.14	5.36		6.50
3005	4", vertical bracket box, 1-1/2" with						
3010	RMX clamps	EA	0.145	11.75	13.25		25.00
3020	BX clamps	"	0.145	12.75	13.25		26.00
3025	4", octagon device cover						
3030	1 switch	EA	0.059	6.91	5.36		12.25
3040	1 duplex recept	"	0.059	6.91	5.36		12.25
3050	4", octagon swivel hanger box, 1/2" hub	"	0.059	18.50	5.36		23.75
3060	3/4" hub	"	0.059	21.00	5.36		26.25
3065	4" octagon adjustable bar hangers						

CONDUCTORS, CONDUIT AND RACEWAYS

ID Code	Component Descriptions	Unit of Meas.	Manhr / Unit	Material Cost	Labor Cost	Equipment Cost	Total Cost
	Description	**Output**		**Unit Costs**			

26 - 05339   PULL AND JUNCTION BOXES, Cont'd...   26 - 05339

ID Code	Component Descriptions	Unit of Meas.	Manhr / Unit	Material Cost	Labor Cost	Equipment Cost	Total Cost
3070	18-1/2"	EA	0.050	8.58	4.52		13.00
3080	26-1/2"	"	0.050	9.37	4.52		14.00
3085	With clip						
3090	18-1/2"	EA	0.050	6.34	4.52		10.75
3100	26-1/2"	"	0.050	7.12	4.52		11.75
3105	4", square face bracket boxes, 1-1/2"						
3110	RMX	EA	0.145	14.00	13.25		27.25
3120	BX	"	0.145	15.25	13.25		28.50
3130	4" square to round plaster rings	"	0.059	4.67	5.36		10.00
3140	2 gang device plaster rings	"	0.059	4.82	5.36		10.25
3145	Surface covers						
3150	1 gang switch	EA	0.059	4.21	5.36		9.57
3160	2 gang switch	"	0.059	4.31	5.36		9.67
3170	1 single recept	"	0.059	6.34	5.36		11.75
3180	1 20a twist lock recept	"	0.059	7.93	5.36		13.25
3190	1 30a twist lock recept	"	0.059	10.25	5.36		15.50
3200	1 duplex recept	"	0.059	3.93	5.36		9.29
3210	2 duplex recept	"	0.059	3.93	5.36		9.29
3220	Switch and duplex recept	"	0.059	6.56	5.36		12.00
3325	4" plastic round boxes, ground straps						
3330	Box only	EA	0.145	2.99	13.25		16.25
3340	Box w/clamps	"	0.200	3.48	18.00		21.50
3350	Box w/16" bar	"	0.229	7.39	20.75		28.25
3360	Box w/24" bar	"	0.250	7.37	22.50		29.75
3370	4" plastic round box covers						
3380	Blank cover	EA	0.059	1.95	5.36		7.31
3390	Plaster ring	"	0.059	3.19	5.36		8.55
3395	4" plastic square boxes						
3400	Box only	EA	0.145	2.31	13.25		15.50
3410	Box w/clamps	"	0.200	2.86	18.00		20.75
3420	Box w/hanger	"	0.250	3.52	22.50		26.00
3430	Box w/nails and clamp	"	0.250	5.04	22.50		27.50
3435	4" plastic square box covers						
3440	Blank cover	EA	0.059	1.90	5.36		7.26
3450	1 gang ring	"	0.059	2.31	5.36		7.67
3460	2 gang ring	"	0.059	3.24	5.36		8.60
3470	Round ring	"	0.059	2.57	5.36		7.93

CONDUCTORS, CONDUIT AND RACEWAYS

ID Code	Description Component Descriptions	Output Unit of Meas.	Manhr / Unit	Material Cost	Labor Cost	Equipment Cost	Total Cost
26 - 05341	**ALUMINUM CONDUIT**						**26 - 05341**
1010	Aluminum conduit						
1020	1/2"	LF	0.030	3.29	2.73		6.02
1040	3/4"	"	0.040	4.22	3.61		7.83
1060	1"	"	0.050	5.92	4.52		10.50
1490	90 deg. elbow						
1500	1/2"	EA	0.190	25.25	17.25		42.50
1510	3/4"	"	0.250	34.50	22.50		57.00
1520	1"	"	0.308	48.00	27.75		76.00
1980	Coupling						
2000	1/2"	EA	0.050	5.17	4.52		9.69
2020	3/4"	"	0.059	7.83	5.36		13.25
2040	1"	"	0.080	10.25	7.23		17.50
26 - 05342	**EMT CONDUIT**						**26 - 05342**
0080	EMT conduit						
0100	1/2"	LF	0.030	0.76	2.73		3.49
1020	3/4"	"	0.040	1.40	3.61		5.01
1030	1"	"	0.050	2.33	4.52		6.85
2980	90 deg. elbow						
3000	1/2"	EA	0.089	7.20	8.04		15.25
3040	3/4"	"	0.100	7.90	9.04		17.00
3060	1"	"	0.107	12.25	9.64		22.00
3980	Connector, steel compression						
4000	1/2"	EA	0.089	2.20	8.04		10.25
4040	3/4"	"	0.089	4.21	8.04		12.25
4060	1"	"	0.089	6.35	8.04		14.50
4480	Coupling, steel, compression						
4500	1/2"	EA	0.059	3.75	5.36		9.11
4540	3/4"	"	0.059	5.10	5.36		10.50
4560	1"	"	0.059	7.71	5.36		13.00
4980	1 hole strap, steel						
5000	1/2"	EA	0.040	0.25	3.61		3.86
5040	3/4"	"	0.040	0.30	3.61		3.91
5060	1"	"	0.040	0.47	3.61		4.08
6000	Connector, steel set screw						
6010	1/2"	EA	0.070	1.68	6.29		7.97
6020	3/4"	"	0.070	2.69	6.29		8.98
6030	1"	"	0.070	4.65	6.29		11.00

CONDUCTORS, CONDUIT AND RACEWAYS

	Description	Output		Unit Costs			
ID Code	Component Descriptions	Unit of Meas.	Manhr / Unit	Material Cost	Labor Cost	Equipment Cost	Total Cost
26 - 05342		**EMT CONDUIT, Cont'd...**					**26 - 05342**
6105	Insulated throat						
6110	1/2"	EA	0.070	2.23	6.29		8.52
6120	3/4"	"	0.070	3.62	6.29		9.91
6130	1"	"	0.070	5.98	6.29		12.25
6205	Connector, die cast set screw						
6210	1/2"	EA	0.059	1.15	5.36		6.51
6220	3/4"	"	0.059	1.96	5.36		7.32
6230	1"	"	0.059	3.69	5.36		9.05
6305	Insulated throat						
6310	1/2"	EA	0.059	2.45	5.36		7.81
6320	3/4"	"	0.059	3.96	5.36		9.32
6330	1"	"	0.059	6.90	5.36		12.25
6405	Coupling, steel set screw						
6410	1/2"	EA	0.040	3.03	3.61		6.64
6420	3/4"	"	0.040	4.56	3.61		8.17
6430	1"	"	0.040	7.40	3.61		11.00
6505	Diecast set screw						
6510	1/2"	EA	0.040	1.06	3.61		4.67
6520	3/4"	"	0.040	1.71	3.61		5.32
6530	1"	"	0.040	2.83	3.61		6.44
6605	1 hole malleable straps						
6610	1/2"	EA	0.040	0.49	3.61		4.10
6620	3/4"	"	0.040	0.67	3.61		4.28
6630	1"	"	0.040	1.11	3.61		4.72
6705	EMT to rigid compression coupling						
6710	1/2"	EA	0.100	5.72	9.04		14.75
6720	3/4"	"	0.100	8.16	9.04		17.25
6730	1"	"	0.150	12.50	13.50		26.00
6735	Set screw couplings						
6740	1/2"	EA	0.100	1.50	9.04		10.50
6750	3/4"	"	0.100	2.26	9.04		11.25
6760	1"	"	0.145	3.78	13.25		17.00
6765	Set screw offset connectors						
6770	1/2"	EA	0.100	3.35	9.04		12.50
6780	3/4"	"	0.100	4.51	9.04		13.50
6790	1"	"	0.145	8.17	13.25		21.50
6795	Compression offset connectors						
6800	1/2"	EA	0.100	5.53	9.04		14.50

CONDUCTORS, CONDUIT AND RACEWAYS

ID Code	Description — Component Descriptions	Output — Unit of Meas.	Output — Manhr / Unit	Unit Costs — Material Cost	Unit Costs — Labor Cost	Unit Costs — Equipment Cost	Unit Costs — Total Cost
26 - 05342	**EMT CONDUIT, Cont'd...**						**26 - 05342**
6810	3/4"	EA	0.100	7.01	9.04		16.00
6820	1"	"	0.145	10.25	13.25		23.50
6825	Type LB set screw condulets						
6830	1/2"	EA	0.229	15.75	20.75		36.50
6840	3/4"	"	0.296	19.25	26.75		46.00
6850	1"	"	0.381	29.25	34.50		64.00
6925	Type T set screw condulets						
6930	1/2"	EA	0.296	19.75	26.75		46.50
6940	3/4"	"	0.400	24.75	36.25		61.00
6950	1"	"	0.444	35.75	40.25		76.00
6985	Type C set screw condulets						
6990	1/2"	EA	0.250	16.50	22.50		39.00
7000	3/4"	"	0.296	20.75	26.75		47.50
7010	1"	"	0.381	30.75	34.50		65.00
7045	Type LL set screw condulets						
7050	1/2"	EA	0.250	16.50	22.50		39.00
7060	3/4"	"	0.296	20.25	26.75		47.00
7070	1"	"	0.381	30.50	34.50		65.00
7105	Type LR set screw condulets						
7110	1/2"	EA	0.250	16.50	22.50		39.00
7120	3/4"	"	0.296	20.25	26.75		47.00
7130	1"	"	0.381	30.50	34.50		65.00
7165	Type LB compression condulets						
7170	1/2"	EA	0.296	37.25	26.75		64.00
7180	3/4"	"	0.500	55.00	45.25		100
7190	1"	"	0.500	70.00	45.25		120
7195	Type T compression condulets						
7200	1/2"	EA	0.400	49.75	36.25		86.00
7210	3/4"	"	0.444	66.00	40.25		110
7220	1"	"	0.615	100	56.00		160
7225	Condulet covers						
7230	1/2"	EA	0.123	2.29	11.25		13.50
7240	3/4"	"	0.123	2.78	11.25		14.00
7250	1"	"	0.123	3.79	11.25		15.00
7325	Clamp type entrance caps						
7330	1/2"	EA	0.250	12.25	22.50		34.75
7340	3/4"	"	0.296	14.50	26.75		41.25
7350	1"	"	0.400	17.00	36.25		53.00

CONDUCTORS, CONDUIT AND RACEWAYS

ID Code	Component Descriptions	Unit of Meas.	Manhr / Unit	Material Cost	Labor Cost	Equipment Cost	Total Cost
	Description	**Output**		**Unit Costs**			
26 - 05342	**EMT CONDUIT, Cont'd...**						**26 - 05342**
7425	Slip fitter type entrance caps						
7430	1/2"	EA	0.250	8.96	22.50		31.50
7440	3/4"	"	0.296	10.75	26.75		37.50
7450	1"	"	0.400	13.00	36.25		49.25
26 - 05343	**FLEXIBLE CONDUIT**						**26 - 05343**
0080	Flexible conduit, steel						
0100	3/8"	LF	0.030	0.96	2.73		3.69
1020	1/2	"	0.030	1.09	2.73		3.82
1040	3/4"	"	0.040	1.50	3.61		5.11
1060	1"	"	0.040	2.85	3.61		6.46
1200	Flexible conduit, liquid tight						
1210	3/8"	LF	0.030	2.68	2.73		5.41
1220	1/2"	"	0.030	3.03	2.73		5.76
1230	3/4"	"	0.040	4.13	3.61		7.74
1240	1"	"	0.040	6.22	3.61		9.83
2000	Connector, straight						
2020	3/8"	EA	0.080	4.52	7.23		11.75
2040	1/2"	"	0.080	4.84	7.23		12.00
2060	3/4"	"	0.089	6.16	8.04		14.25
2080	1"	"	0.100	11.00	9.04		20.00
4380	Straight insulated throat connectors						
4400	3/8"	EA	0.123	5.49	11.25		16.75
4440	1/2"	"	0.123	5.49	11.25		16.75
4450	3/4"	"	0.145	8.03	13.25		21.25
4460	1"	"	0.145	12.50	13.25		25.75
4525	90 deg connectors						
4530	3/8"	EA	0.148	7.67	13.50		21.25
4540	1/2"	"	0.148	7.67	13.50		21.25
4550	3/4"	"	0.170	12.25	15.50		27.75
4560	1"	"	0.182	23.75	16.50		40.25
4625	90 degree insulated throat connectors						
4630	3/8"	EA	0.145	9.48	13.25		22.75
4640	1/2"	"	0.145	9.48	13.25		22.75
4650	3/4"	"	0.170	14.25	15.50		29.75
4660	1"	"	0.178	27.25	16.00		43.25
4800	Flexible aluminum conduit						
4810	3/8"	LF	0.030	0.56	2.73		3.29

CONDUCTORS, CONDUIT AND RACEWAYS

ID Code	Description / Component Descriptions	Output Unit of Meas.	Output Manhr / Unit	Unit Costs Material Cost	Unit Costs Labor Cost	Unit Costs Equipment Cost	Unit Costs Total Cost
26 - 05343	**FLEXIBLE CONDUIT, Cont'd...**						**26 - 05343**
4820	1/2"	LF	0.030	0.67	2.73		3.40
4830	3/4"	"	0.040	0.92	3.61		4.53
4840	1"	"	0.040	1.74	3.61		5.35
5000	Connector, straight						
5020	3/8"	EA	0.100	1.72	9.04		10.75
5040	1/2"	"	0.100	2.40	9.04		11.50
5060	3/4"	"	0.107	2.62	9.64		12.25
5080	1"	"	0.123	9.61	11.25		20.75
6115	Straight insulated throat connectors						
6120	3/8"	EA	0.089	1.68	8.04		9.72
6130	1/2"	"	0.089	3.38	8.04		11.50
6140	3/4"	"	0.089	3.59	8.04		11.75
6150	1"	"	0.100	8.74	9.04		17.75
6225	90 deg connectors						
6230	3/8"	EA	0.145	2.76	13.25		16.00
6240	1/2"	"	0.145	4.66	13.25		18.00
6250	3/4"	"	0.145	7.32	13.25		20.50
6260	1"	"	0.170	12.75	15.50		28.25
6315	90 deg insulated throat connectors						
6320	3/8"	EA	0.145	3.42	13.25		16.75
6330	1/2"	"	0.145	5.18	13.25		18.50
6340	3/4"	"	0.145	8.64	13.25		22.00
6350	1"	"	0.170	14.25	15.50		29.75
26 - 05344	**GALVANIZED CONDUIT**						**26 - 05344**
1980	Galvanized rigid steel conduit						
2000	1/2"	LF	0.040	3.89	3.61		7.50
2040	3/4"	"	0.050	4.30	4.52		8.82
2060	1"	"	0.059	6.21	5.36		11.50
2080	1-1/4"	"	0.080	8.59	7.23		15.75
2100	1-1/2"	"	0.089	10.00	8.04		18.00
2120	2"	"	0.100	12.75	9.04		21.75
2480	90 degree ell						
2500	1/2"	EA	0.250	10.25	22.50		32.75
2540	3/4"	"	0.308	10.75	27.75		38.50
2560	1"	"	0.381	16.50	34.50		51.00
2580	1-1/4"	"	0.444	22.75	40.25		63.00
2590	1-1/2"	"	0.500	28.00	45.25		73.00

CONDUCTORS, CONDUIT AND RACEWAYS

ID Code	Description / Component Descriptions	Unit of Meas.	Manhr / Unit	Material Cost	Labor Cost	Equipment Cost	Total Cost
26 - 05344	**GALVANIZED CONDUIT, Cont'd...**						**26 - 05344**
2600	2"	EA	0.533	40.75	48.25		89.00
3200	Couplings, with set screws						
3220	1/2"	EA	0.050	5.17	4.52		9.69
3260	3/4"	"	0.059	6.82	5.36		12.25
3280	1"	"	0.080	11.00	7.23		18.25
3300	1-1/4"	"	0.100	18.50	9.04		27.50
3320	1-1/2"	"	0.123	24.00	11.25		35.25
3340	2"	"	0.145	54.00	13.25		67.00
3440	Split couplings						
3460	1/2"	EA	0.190	4.40	17.25		21.75
3480	3/4"	"	0.250	5.72	22.50		28.25
3500	1"	"	0.276	8.03	25.00		33.00
3520	1-1/4"	"	0.308	15.75	27.75		43.50
3540	1-1/2"	"	0.381	20.50	34.50		55.00
3560	2"	"	0.571	47.25	52.00		99.00
3780	Erickson couplings						
3800	1/2"	EA	0.444	6.29	40.25		46.50
3840	3/4"	"	0.500	7.69	45.25		53.00
3850	1"	"	0.615	13.50	56.00		70.00
3860	1-1/4"	"	0.889	24.25	80.00		100
3870	1-1/2"	"	1.000	31.50	90.00		120
3880	2"	"	1.333	61.00	120		180
4000	Seal fittings						
4020	1/2"	EA	0.667	20.75	60.00		81.00
4040	3/4"	"	0.800	22.75	72.00		95.00
4060	1"	"	1.000	29.00	90.00		120
4080	1-1/4"	"	1.143	30.00	100		130
4100	1-1/2"	"	1.333	53.00	120		170
4120	2"	"	1.600	58.00	140		200
4980	Entrance fitting (weatherhead), threaded						
5000	1/2"	EA	0.444	9.63	40.25		50.00
5040	3/4"	"	0.500	11.75	45.25		57.00
5060	1"	"	0.571	15.25	52.00		67.00
5080	1-1/4"	"	0.727	19.75	66.00		86.00
5100	1-1/2"	"	0.800	34.75	72.00		110
5120	2"	"	0.889	53.00	80.00		130
5980	Locknuts						
6000	1/2"	EA	0.050	0.24	4.52		4.76

CONDUCTORS, CONDUIT AND RACEWAYS

ID Code	Description		Output		Unit Costs			
	Component Descriptions	Unit of Meas.	Manhr / Unit	Material Cost	Labor Cost	Equipment Cost	Total Cost
26 - 05344	**GALVANIZED CONDUIT, Cont'd...**						**26 - 05344**
6040	3/4"	EA	0.050	0.31	4.52		4.83
6060	1"	"	0.050	0.49	4.52		5.01
6080	1-1/4"	"	0.050	0.67	4.52		5.19
6100	1-1/2"	"	0.059	1.12	5.36		6.48
6120	2"	"	0.059	1.62	5.36		6.98
6250	Plastic conduit bushings						
6260	1/2"	EA	0.123	0.49	11.25		11.75
6300	3/4"	"	0.145	0.74	13.25		14.00
6320	1"	"	0.190	1.05	17.25		18.25
6340	1-1/4"	"	0.222	1.38	20.00		21.50
6360	1-1/2"	"	0.250	1.87	22.50		24.25
6380	2"	"	0.308	4.25	27.75		32.00
6660	Conduit bushings, steel						
6680	1/2"	EA	0.123	0.68	11.25		12.00
6720	3/4"	"	0.145	0.86	13.25		14.00
6740	1"	"	0.190	1.31	17.25		18.50
6760	1-1/4"	"	0.222	1.87	20.00		21.75
6780	1-1/2"	"	0.250	2.67	22.50		25.25
6800	2"	"	0.308	5.43	27.75		33.25
7500	Pipe cap						
7520	1/2"	EA	0.050	0.63	4.52		5.15
7540	3/4"	"	0.050	0.69	4.52		5.21
7560	1"	"	0.050	1.10	4.52		5.62
7580	1-1/4"	"	0.080	1.90	7.23		9.13
7600	1-1/2"	"	0.080	2.95	7.23		10.25
7620	2"	"	0.080	3.32	7.23		10.50
8215	Threaded couplings						
8220	1/2"	EA	0.050	2.37	4.52		6.89
8230	3/4"	"	0.059	2.91	5.36		8.27
8240	1"	"	0.080	4.32	7.23		11.50
8250	1-1/4"	"	0.089	5.40	8.04		13.50
8260	1-1/2"	"	0.100	6.63	9.04		15.75
8270	2"	"	0.107	9.02	9.64		18.75
8335	Threadless couplings						
8340	1/2"	EA	0.100	9.65	9.04		18.75
8350	3/4"	"	0.123	10.00	11.25		21.25
8360	1"	"	0.145	13.00	13.25		26.25
8370	1-1/4"	"	0.190	15.25	17.25		32.50

CONDUCTORS, CONDUIT AND RACEWAYS

ID Code	Component Descriptions	Unit of Meas.	Manhr / Unit	Material Cost	Labor Cost	Equipment Cost	Total Cost
26 - 05344	**GALVANIZED CONDUIT, Cont'd...**						26 - 05344
8380	1-1/2"	EA	0.250	18.00	22.50		40.50
8390	2"	"	0.308	26.50	27.75		54.00
8455	Threadless connectors						
8460	1/2"	EA	0.100	4.59	9.04		13.75
8470	3/4"	"	0.123	7.35	11.25		18.50
8480	1"	"	0.145	11.50	13.25		24.75
8490	1-1/4"	"	0.190	19.75	17.25		37.00
8500	1-1/2"	"	0.250	30.00	22.50		53.00
8510	2"	"	0.308	58.00	27.75		86.00
8575	Setscrew connectors						
8580	1/2"	EA	0.080	3.35	7.23		10.50
8590	3/4"	"	0.089	4.65	8.04		12.75
8600	1"	"	0.100	7.25	9.04		16.25
8610	1-1/4"	"	0.123	12.75	11.25		24.00
8620	1-1/2"	"	0.145	18.50	13.25		31.75
8630	2"	"	0.190	36.75	17.25		54.00
8695	Clamp type entrance caps						
8700	1/2"	EA	0.308	13.50	27.75		41.25
8710	3/4"	"	0.381	16.00	34.50		51.00
8720	1"	"	0.444	22.25	40.25		63.00
8730	1-1/4"	"	0.500	26.00	45.25		71.00
8740	1-1/2"	"	0.615	47.00	56.00		100
8750	2"	"	0.727	57.00	66.00		120
8795	LB condulets						
8800	1/2"	EA	0.308	16.00	27.75		43.75
8810	3/4"	"	0.381	19.50	34.50		54.00
8820	1"	"	0.444	29.25	40.25		70.00
8830	1-1/4"	"	0.500	50.00	45.25		95.00
8840	1-1/2"	"	0.615	66.00	56.00		120
8850	2"	"	0.727	110	66.00		180
8895	T condulets						
8900	1/2"	EA	0.381	20.25	34.50		55.00
8910	3/4"	"	0.444	24.25	40.25		65.00
8920	1"	"	0.500	36.50	45.25		82.00
8930	1-1/4"	"	0.571	53.00	52.00		110
8940	1-1/2"	"	0.615	71.00	56.00		130
8950	2"	"	0.727	110	66.00		180
8995	X condulets						

CONDUCTORS, CONDUIT AND RACEWAYS

ID Code	Component Descriptions	Unit of Meas.	Manhr / Unit	Material Cost	Labor Cost	Equipment Cost	Total Cost
26 - 05344	**GALVANIZED CONDUIT, Cont'd...**						**26 - 05344**
9000	1/2"	EA	0.444	29.75	40.25		70.00
9010	3/4"	"	0.500	32.25	45.25		78.00
9020	1"	"	0.571	53.00	52.00		110
9030	1-1/4"	"	0.615	69.00	56.00		130
9040	1-1/2"	"	0.667	89.00	60.00		150
9050	2"	"	0.879	180	80.00		260
9055	Blank steel condulet covers						
9060	1/2"	EA	0.100	4.59	9.04		13.75
9070	3/4"	"	0.100	5.68	9.04		14.75
9080	1"	"	0.100	7.73	9.04		16.75
9090	1-1/4"	"	0.123	9.47	11.25		20.75
9100	1-1/2"	"	0.123	9.94	11.25		21.25
9110	2"	"	0.123	16.75	11.25		28.00
9155	Solid condulet gaskets						
9160	1/2"	EA	0.050	3.78	4.52		8.30
9170	3/4"	"	0.050	4.09	4.52		8.61
9180	1"	"	0.050	4.73	4.52		9.25
9190	1-1/4"	"	0.080	5.88	7.23		13.00
9200	1-1/2"	"	0.080	6.19	7.23		13.50
9210	2"	"	0.080	6.94	7.23		14.25
9255	One-hole malleable straps						
9260	1/2"	EA	0.040	0.60	3.61		4.21
9270	3/4"	"	0.040	0.83	3.61		4.44
9280	1"	"	0.040	1.20	3.61		4.81
9290	1-1/4"	"	0.050	2.40	4.52		6.92
9300	1-1/2"	"	0.050	2.76	4.52		7.28
9310	2"	"	0.050	5.41	4.52		9.93
9375	One-hole steel straps						
9380	1/2"	EA	0.040	0.15	3.61		3.76
9390	3/4"	"	0.040	0.22	3.61		3.83
9400	1"	"	0.040	0.38	3.61		3.99
9410	1-1/4"	"	0.050	0.54	4.52		5.06
9412	1-1/2"	"	0.050	0.73	4.52		5.25
9413	2"	"	0.050	0.95	4.52		5.47
9650	Grounding locknuts						
9670	1/2"	EA	0.080	3.64	7.23		10.75
9680	3/4"	"	0.080	4.59	7.23		11.75
9690	1"	"	0.080	6.64	7.23		13.75

CONDUCTORS, CONDUIT AND RACEWAYS

ID Code	Component Descriptions	Unit of Meas.	Manhr / Unit	Material Cost	Labor Cost	Equipment Cost	Total Cost
26 - 05344	**GALVANIZED CONDUIT, Cont'd...**						**26 - 05344**
9700	1-1/4"	EA	0.089	7.12	8.04		15.25
9710	1-1/2"	"	0.089	7.43	8.04		15.50
9720	2"	"	0.089	11.00	8.04		19.00
9758	Insulated grounding metal bushings						
9760	1/2"	EA	0.190	2.58	17.25		19.75
9765	3/4"	"	0.222	3.81	20.00		23.75
9770	1"	"	0.250	5.44	22.50		28.00
9775	1-1/4"	"	0.308	8.67	27.75		36.50
9780	1-1/2"	"	0.381	10.75	34.50		45.25
9785	2"	"	0.444	15.75	40.25		56.00
26 - 05345	**PLASTIC COATED CONDUIT**						**26 - 05345**
0980	Rigid steel conduit, plastic coated						
1000	1/2"	LF	0.050	9.52	4.52		14.00
1040	3/4"	"	0.059	11.00	5.36		16.25
1060	1"	"	0.080	14.25	7.23		21.50
1080	1-1/4"	"	0.100	18.00	9.04		27.00
1100	1-1/2"	"	0.123	22.00	11.25		33.25
1120	2"	"	0.145	28.75	13.25		42.00
1490	90 degree elbows						
1500	1/2"	EA	0.308	38.25	27.75		66.00
1540	3/4"	"	0.381	39.75	34.50		74.00
1560	1"	"	0.444	45.50	40.25		86.00
1580	1-1/4"	"	0.500	56.00	45.25		100
1600	1-1/2"	"	0.615	69.00	56.00		130
1620	2"	"	0.800	96.00	72.00		170
1980	Couplings						
2000	1/2"	EA	0.059	11.00	5.36		16.25
2040	3/4"	"	0.080	11.50	7.23		18.75
2060	1"	"	0.089	15.25	8.04		23.25
2080	1-1/4"	"	0.107	17.75	9.64		27.50
2100	1-1/2"	"	0.123	24.75	11.25		36.00
2120	2"	"	0.145	31.00	13.25		44.25
2980	1 hole conduit straps						
3000	3/4"	EA	0.050	17.00	4.52		21.50
3010	1"	"	0.050	17.50	4.52		22.00
3020	1-1/4"	"	0.059	25.50	5.36		30.75
3030	1-1/2"	"	0.059	27.00	5.36		32.25

CONDUCTORS, CONDUIT AND RACEWAYS

ID Code	Description Component Descriptions	Output Unit of Meas.	Output Manhr / Unit	Unit Costs Material Cost	Unit Costs Labor Cost	Unit Costs Equipment Cost	Unit Costs Total Cost
26 - 05345	**PLASTIC COATED CONDUIT, Cont'd...**						**26 - 05345**
3040	2"	EA	0.059	39.25	5.36		44.50
26 - 05346	**PLASTIC CONDUIT**						**26 - 05346**
3010	PVC conduit, schedule 40						
3020	1/2"	LF	0.030	1.01	2.73		3.74
3040	3/4"	"	0.030	1.25	2.73		3.98
3060	1"	"	0.040	1.81	3.61		5.42
3080	1-1/4"	"	0.040	2.51	3.61		6.12
3100	1-1/2"	"	0.050	3.00	4.52		7.52
3120	2"	"	0.050	3.58	4.52		8.10
3480	Couplings						
3500	1/2"	EA	0.050	0.62	4.52		5.14
3520	3/4"	"	0.050	0.74	4.52		5.26
3540	1"	"	0.050	1.16	4.52		5.68
3560	1-1/4"	"	0.059	1.53	5.36		6.89
3580	1-1/2"	"	0.059	2.12	5.36		7.48
3600	2"	"	0.059	2.78	5.36		8.14
3705	90 degree elbows						
3710	1/2"	EA	0.100	2.40	9.04		11.50
3740	3/4"	"	0.123	2.62	11.25		13.75
3760	1"	"	0.123	4.15	11.25		15.50
3780	1-1/4"	"	0.145	5.78	13.25		19.00
3800	1-1/2"	"	0.190	7.85	17.25		25.00
3810	2"	"	0.222	11.00	20.00		31.00
3900	Terminal adapters						
4000	1/2"	EA	0.100	0.90	9.04		9.94
4040	3/4"	"	0.100	1.47	9.04		10.50
4060	1"	"	0.100	1.83	9.04		10.75
4080	1-1/4"	"	0.160	2.30	14.50		16.75
4100	1-1/2"	"	0.160	2.94	14.50		17.50
4120	2"	"	0.160	4.06	14.50		18.50
5010	LB conduit body						
5020	1/2"	EA	0.190	7.86	17.25		25.00
5040	3/4"	"	0.190	10.25	17.25		27.50
5060	1	"	0.190	11.25	17.25		28.50
5080	1-1/4"	"	0.308	17.00	27.75		44.75
5100	1-1/2"	"	0.308	20.50	27.75		48.25
5120	2"	"	0.308	36.00	27.75		64.00

CONDUCTORS, CONDUIT AND RACEWAYS

ID Code	Component Descriptions	Unit of Meas.	Manhr / Unit	Material Cost	Labor Cost	Equipment Cost	Total Cost
	Description	**Output**		**Unit Costs**			
26 - 05346	**PLASTIC CONDUIT, Cont'd...**						**26 - 05346**
7485	PVC cement						
7490	1 pint	EA					20.50
7500	1 quart	"					31.75
26 - 05347	**STEEL CONDUIT**						**26 - 05347**
7980	Intermediate metal conduit (IMC)						
8000	1/2"	LF	0.030	3.15	2.73		5.88
8040	3/4"	"	0.040	3.87	3.61		7.48
8060	1"	"	0.050	5.86	4.52		10.50
8080	1-1/4"	"	0.059	7.50	5.36		12.75
8100	1-1/2"	"	0.080	9.39	7.23		16.50
8120	2"	"	0.089	12.25	8.04		20.25
8490	90 degree ell						
8500	1/2"	EA	0.250	24.25	22.50		46.75
8540	3/4"	"	0.308	25.25	27.75		53.00
8560	1"	"	0.381	38.75	34.50		73.00
8580	1-1/4"	"	0.444	54.00	40.25		94.00
8600	1-1/2"	"	0.500	66.00	45.25		110
8620	2"	"	0.571	95.00	52.00		150
9260	Couplings						
9280	1/2"	EA	0.050	5.91	4.52		10.50
9290	3/4"	"	0.059	7.26	5.36		12.50
9300	1"	"	0.080	10.75	7.23		18.00
9310	1-1/4"	"	0.089	13.50	8.04		21.50
9320	1-1/2"	"	0.100	17.00	9.04		26.00
9330	2"	"	0.107	22.50	9.64		32.25

SERVICE AND DISTRIBUTION

ID Code	Component Descriptions	Unit of Meas.	Manhr / Unit	Material Cost	Labor Cost	Equipment Cost	Total Cost
26 - 24160	**PANELBOARDS**						**26 - 24160**
1000	Indoor load center, 1 phase 240v main lug only						
1020	30a - 2 spaces	EA	2.000	41.75	180		220
1030	100a - 8 spaces	"	2.424	130	220		350
1040	150a - 16 spaces	"	2.963	350	270		620
1050	200a - 24 spaces	"	3.478	720	310		1,030
1060	200a - 42 spaces	"	4.000	740	360		1,100
1100	Main circuit breaker						
1110	100a - 8 spaces	EA	2.424	420	220		640
1120	100a - 16 spaces	"	2.759	460	250		710

SERVICE AND DISTRIBUTION

ID Code	Description		Output		Unit Costs			
	Component Descriptions	Unit of Meas.	Manhr / Unit	Material Cost	Labor Cost	Equipment Cost	Total Cost
26 - 24160	**PANELBOARDS, Cont'd...**						**26 - 24160**	
1130	150a - 16 spaces	EA	2.963	750	270		1,020
1140	150a - 24 spaces	"	3.200	890	290		1,180
1150	200a - 24 spaces	"	3.478	830	310		1,140
1160	200a - 42 spaces	"	3.636	1,190	330		1,520
2510	120/208v, flush, 3 ph., 4 wire, main only						
2515	100a						
2520	12 circuits	EA	5.096	1,320	460		1,780
2540	20 circuits	"	6.299	1,820	570		2,390
2560	30 circuits	"	7.018	2,710	630		3,340
2570	225a						
2580	30 circuits	EA	7.767	2,760	700		3,460
2600	42 circuits	"	9.524	3,490	860		4,350

BASIC MATERIALS

ID Code	Description		Output		Unit Costs			
26 - 25003	**BUS DUCT**						**26 - 25003**	
1000	Bus duct, 100a, plug-in						
1010	10', 600v	EA	2.759	560	250		810
1020	With ground	"	4.211	750	380		1,130
1145	Circuit breakers, with enclosure						
1147	1 pole						
1150	15a-60a	EA	1.000	540	90.00		630
1160	70a-100a	"	1.250	620	110		730
1165	2 pole						
1170	15a-60a	EA	1.100	760	100		860
1180	70a-100a	"	1.301	910	120		1,030
1985	Circuit breaker, adapter cubicle						
1990	225a	EA	1.509	8,280	140		8,420
2000	400a	"	1.600	9,770	140		9,910
3075	Fusible switches, 240v, 3 phase						
3080	30a	EA	1.000	730	90.00		820
3090	60a	"	1.250	890	110		1,000
3100	100a	"	1.509	1,190	140		1,330
3110	200a	"	2.105	2,080	190		2,270

BASIC MATERIALS

	Description		Output		Unit Costs			
ID Code	Component Descriptions	Unit of Meas.	Manhr / Unit	Material Cost	Labor Cost	Equipment Cost	Total Cost
26 - 27130	**METERING**						**26 - 27130**
0490	Outdoor wp meter sockets, 1 gang, 240v, 1 phase						
0510	Includes sealing ring, 100a	EA	1.509	76.00	140		220
0520	150a	"	1.778	90.00	160		250
0530	200a	"	2.000	110	180		290
0570	Die cast hubs, 1-1/4"	"	0.320	10.50	29.00		39.50
0580	1-1/2"	"	0.320	12.00	29.00		41.00
0590	2"	"	0.320	14.50	29.00		43.50
26 - 27268	**RECEPTACLES**						**26 - 27268**
0490	Contractor grade duplex receptacles, 15a 120v						
0510	Duplex	EA	0.200	2.35	18.00		20.25
1000	125 volt, 20a, duplex, standard grade	"	0.200	17.50	18.00		35.50
1040	Ground fault interrupter type	"	0.296	57.00	26.75		84.00
1520	250 volt, 20a, 2 pole, single, grounding type	"	0.200	29.25	18.00		47.25
1540	120/208v, 4 pole, single receptacle, twist lock						
1560	20a	EA	0.348	34.75	31.50		66.00
1580	50a	"	0.348	66.00	31.50		98.00
1590	125/250v, 3 pole, flush receptacle						
1600	30a	EA	0.296	35.25	26.75		62.00
1620	50a	"	0.296	43.50	26.75		70.00
1640	60a	"	0.348	110	31.50		140
2020	Dryer receptacle, 250v, 30a/50a, 3 wire	"	0.296	26.25	26.75		53.00
2040	Clock receptacle, 2 pole, grounding type	"	0.200	17.50	18.00		35.50
3000	125v, 20a single recept. grounding type						
3010	Standard grade	EA	0.200	19.00	18.00		37.00
3105	125/250v, 3 pole, 3 wire surface recepts						
3110	30a	EA	0.296	29.75	26.75		57.00
3120	50a	"	0.296	33.00	26.75		60.00
3130	60a	"	0.348	73.00	31.50		100
3135	Cord set, 3 wire, 6' cord						
3140	30a	EA	0.296	26.50	26.75		53.00
3150	50a	"	0.296	37.50	26.75		64.00
3155	125/250v, 3 pole, 3 wire cap						
3160	30a	EA	0.400	26.25	36.25		63.00
3170	50a	"	0.400	47.75	36.25		84.00
3180	60a	"	0.444	61.00	40.25		100

BASIC MATERIALS

ID Code	Description / Component Descriptions	Output		Unit Costs			
		Unit of Meas.	Manhr / Unit	Material Cost	Labor Cost	Equipment Cost	Total Cost
26 - 28130	**FUSES**						**26 - 28130**
1000	Fuse, one-time, 250v						
1010	30a	EA	0.050	4.23	4.52		8.75
1020	60a	"	0.050	7.16	4.52		11.75
1040	100a	"	0.050	30.00	4.52		34.50
26 - 28161	**CIRCUIT BREAKERS**						**26 - 28161**
5000	Load center circuit breakers, 240v						
5010	1 pole, 10-60a	EA	0.250	32.50	22.50		55.00
5015	2 pole						
5020	10-60a	EA	0.400	66.00	36.25		100
5030	70-100a	"	0.667	200	60.00		260
5040	110-150a	"	0.727	420	66.00		490
5065	Load center, GFI breakers, 240v						
5070	1 pole, 15-30a	EA	0.296	250	26.75		280
5095	Tandem breakers, 240v						
5100	1 pole, 15-30a	EA	0.400	54.00	36.25		90.00
5110	2 pole, 15-30a	"	0.533	98.00	48.25		150
26 - 28164	**SWITCHES**						**26 - 28164**
4000	Photoelectric switches						
4010	1000 watt						
4020	105-135v	EA	0.727	56.00	66.00		120
4970	Dimmer switch and switch plate						
4990	600w	EA	0.308	52.00	27.75		80.00
5065	Time clocks with skip, 40a, 120v						
5070	SPST	EA	0.748	160	68.00		230
5171	Contractor grade wall switch 15a, 120v						
5172	Single pole	EA	0.160	2.73	14.50		17.25
5173	Three way	"	0.200	4.99	18.00		23.00
5174	Four way	"	0.267	16.75	24.00		40.75
5175	Specification grade toggle switches, 20a, 120-277v						
5180	Single pole	EA	0.200	6.01	18.00		24.00
5190	Double pole	"	0.296	14.50	26.75		41.25
5200	3 way	"	0.250	15.75	22.50		38.25
5210	4 way	"	0.296	47.25	26.75		74.00
5440	Combination switch and pilot light, single pole	"	0.296	20.75	26.75		47.50
5450	3 way	"	0.348	25.75	31.50		57.00
5460	Combination switch and receptacle, single pole	"	0.296	29.75	26.75		57.00
5470	3 way	"	0.296	36.50	26.75		63.00

BASIC MATERIALS

ID Code	Description / Component Descriptions	Unit of Meas.	Manhr / Unit	Material Cost	Labor Cost	Equipment Cost	Total Cost
26 - 28164	**SWITCHES, Cont'd...**						**26 - 28164**
5495	Switch plates, plastic ivory						
5510	1 gang	EA	0.080	0.65	7.23		7.88
5520	2 gang	"	0.100	1.54	9.04		10.50
5530	3 gang	"	0.119	2.41	10.75		13.25
5540	4 gang	"	0.145	6.19	13.25		19.50
5550	5 gang	"	0.160	6.48	14.50		21.00
5560	6 gang	"	0.182	7.64	16.50		24.25
5565	Stainless steel						
5570	1 gang	EA	0.080	5.57	7.23		12.75
5580	2 gang	"	0.100	7.73	9.04		16.75
5590	3 gang	"	0.123	12.00	11.25		23.25
5600	4 gang	"	0.145	20.25	13.25		33.50
5610	5 gang	"	0.160	24.00	14.50		38.50
5620	6 gang	"	0.182	29.75	16.50		46.25
5625	Brass						
5630	1 gang	EA	0.080	10.50	7.23		17.75
5640	2 gang	"	0.100	22.25	9.04		31.25
5650	3 gang	"	0.123	34.50	11.25		45.75
5660	4 gang	"	0.145	39.75	13.25		53.00
5670	5 gang	"	0.160	49.25	14.50		64.00
5680	6 gang	"	0.182	59.00	16.50		76.00

FACILITY LIGHTNING PROTECTION

ID Code	Description / Component Descriptions	Unit of Meas.	Manhr / Unit	Material Cost	Labor Cost	Equipment Cost	Total Cost
26 - 41130	**LIGHTNING PROTECTION**						**26 - 41130**
0100	Lightning protection						
0980	Copper point, nickel plated, 12'						
1000	1/2" dia.	EA	1.000	69.00	90.00		160
1020	5/8" dia.	"	1.000	78.00	90.00		170

LIGHTING

ID Code	Description / Component Descriptions	Unit of Meas.	Manhr / Unit	Material Cost	Labor Cost	Equipment Cost	Total Cost
26 - 51101	**INTERIOR LIGHTING**						**26 - 51101**
0010	Recessed fluorescent fixtures, 2'x2'						
0015	2 lamp	EA	0.727	110	66.00		180
0020	4 lamp	"	0.727	150	66.00		220
0205	Surface mounted incandescent fixtures						
0210	40w	EA	0.667	170	60.00		230

LIGHTING

ID Code	Description		Output		Unit Costs			
	Component Descriptions	Unit of Meas.	Manhr / Unit	Material Cost	Labor Cost	Equipment Cost	Total Cost

26 - 51101	**INTERIOR LIGHTING, Cont'd...**						**26 - 51101**
0220	75w	EA	0.667	170	60.00		230
0230	100w	"	0.667	190	60.00		250
0240	150w	"	0.667	250	60.00		310
0245	Pendant						
0250	40w	EA	0.800	140	72.00		210
0260	75w	"	0.800	150	72.00		220
0270	100w	"	0.800	170	72.00		240
0280	150w	"	0.800	190	72.00		260
0281	Contractor grade recessed down lights						
0282	100 watt housing only	EA	1.000	110	90.00		200
0283	150 watt housing only	"	1.000	160	90.00		250
0284	100 watt trim	"	0.500	92.00	45.25		140
0285	150 watt trim	"	0.500	140	45.25		190
0287	Recessed incandescent fixtures						
0290	40w	EA	1.509	230	140		370
0300	75w	"	1.509	250	140		390
0310	100w	"	1.509	270	140		410
0320	150w	"	1.509	290	140		430
0395	Light track single circuit						
0400	2'	EA	0.500	67.00	45.25		110
0410	4'	"	0.500	79.00	45.25		120
0420	8'	"	1.000	110	90.00		200
0430	12'	"	1.509	150	140		290
0435	Fittings and accessories						
0440	Dead end	EA	0.145	26.50	13.25		39.75
0450	Starter kit	"	0.250	35.50	22.50		58.00
0460	Conduit feed	"	0.145	34.50	13.25		47.75
0470	Straight connector	"	0.145	30.50	13.25		43.75
0480	Center feed	"	0.145	48.75	13.25		62.00
0490	L-connector	"	0.145	34.50	13.25		47.75
0501	T-connector	"	0.145	46.25	13.25		60.00
0510	X-connector	"	0.200	56.00	18.00		74.00
0520	Cord and plug	"	0.100	56.00	9.04		65.00
0530	Rigid corner	"	0.145	74.00	13.25		87.00
0540	Flex connector	"	0.145	58.00	13.25		71.00
0550	2 way connector	"	0.200	160	18.00		180
0560	Spacer clip	"	0.050	2.49	4.52		7.01
0570	Grid box	"	0.145	14.00	13.25		27.25

LIGHTING

	Description	Output		Unit Costs			
ID Code	Component Descriptions	Unit of Meas.	Manhr / Unit	Material Cost	Labor Cost	Equipment Cost	Total Cost
26 - 51101	**INTERIOR LIGHTING, Cont'd...**						**26 - 51101**
0580	T-bar clip	EA	0.050	3.72	4.52		8.24
0590	Utility hook	"	0.145	10.75	13.25		24.00
0595	Fixtures, square						
0600	R-20	EA	0.145	68.00	13.25		81.00
0610	R-30	"	0.145	110	13.25		120
0620	40w flood	"	0.145	170	13.25		180
0630	40w spot	"	0.145	170	13.25		180
0640	100w flood	"	0.145	200	13.25		210
0650	100w spot	"	0.145	150	13.25		160
0660	Mini spot	"	0.145	65.00	13.25		78.00
0670	Mini flood	"	0.145	150	13.25		160
0680	Quartz, 500w	"	0.145	380	13.25		390
0690	R-20 sphere	"	0.145	110	13.25		120
0700	R-30 sphere	"	0.145	60.00	13.25		73.00
0710	R-20 cylinder	"	0.145	81.00	13.25		94.00
0720	R-30 cylinder	"	0.145	93.00	13.25		110
0730	R-40 cylinder	"	0.145	95.00	13.25		110
0740	R-30 wall wash	"	0.145	150	13.25		160
0750	R-40 wall wash	"	0.145	180	13.25		190
26 - 51401	**INDUSTRIAL LIGHTING**						**26 - 51401**
0100	Surface mounted fluorescent, wrap around lens						
0110	1 lamp	EA	0.800	140	72.00		210
0120	2 lamps	"	0.889	190	80.00		270
0250	Wall mounted fluorescent						
0300	2-20w lamps	EA	0.500	140	45.25		190
0320	2-30w lamps	"	0.500	160	45.25		210
0340	2-40w lamps	"	0.667	160	60.00		220
0490	Strip fluorescent						
0510	4'						
0520	1 lamp	EA	0.667	69.00	60.00		130
0540	2 lamps	"	0.667	84.00	60.00		140
0550	8'						
0560	1 lamp	EA	0.727	100	66.00		170
0580	2 lamps	"	0.889	150	80.00		230
4700	Compact fluorescent						
4720	2-7w	EA	1.000	240	90.00		330
4740	2-13w	"	1.333	280	120		400

DIVISION 27
COMMUNICATIONS

DISTRIBUTED AUDIO-VIDEO COMMUNICATIONS SYSTEMS

ID Code	Description Component Descriptions	Output		Unit Costs			
		Unit of Meas.	Manhr / Unit	Material Cost	Labor Cost	Equipment Cost	Total Cost
27 - 51002	**SIGNALING SYSTEMS**						**27 - 51002**
2000	Contractor grade doorbell chime kit						
2020	Chime	EA	1.000	55.00	90.00		140
2040	Doorbutton	"	0.320	7.67	29.00		36.75
2050	Transformer	"	0.500	24.50	45.25		70.00

DIVISION 31
EARTHWORK

SITE CLEARING

ID Code	Component Descriptions	Unit of Meas.	Manhr / Unit	Material Cost	Labor Cost	Equipment Cost	Total Cost
31 - 11001	**CLEAR WOODED AREAS**						**31 - 11001**
0980	Clear wooded area						
1000	Light density	ACRE	60.000		4,150	3,940	8,080
1500	Medium density	"	80.000		5,530	5,250	10,780
1800	Heavy density	"	96.000		6,630	6,300	12,930
31 - 13005	**TREE CUTTING & CLEARING**						**31 - 13005**
0980	Cut trees and clear out stumps						
1000	9" to 12" dia.	EA	4.800		330	310	650
1400	To 24" dia.	"	6.000		410	390	810
1600	24" dia. and up	"	8.000		550	530	1,080
5000	Loading and trucking						
5010	For machine load, per load, round trip						
5020	1 mile	EA	0.960		66.00	63.00	130
5025	3 mile	"	1.091		75.00	72.00	150
5030	5 mile	"	1.200		83.00	79.00	160
5035	10 mile	"	1.600		110	110	220
5040	20 mile	"	2.400		170	160	320
5050	Hand loaded, round trip						
5060	1 mile	EA	2.000		140	180	320
5065	3 mile	"	2.286		160	200	360
5070	5 mile	"	2.667		180	240	420
5080	10 mile	"	3.200		220	290	510
5100	20 mile	"	4.000		280	360	630

EARTHWORK, EXCAVATION & FILL

ID Code	Component Descriptions	Unit of Meas.	Manhr / Unit	Material Cost	Labor Cost	Equipment Cost	Total Cost
31 - 22130	**ROUGH GRADING**						**31 - 22130**
1000	Site grading, cut & fill, sandy clay, 200' haul, 75 hp	CY	0.032		2.21	2.86	5.07
1100	Spread topsoil by equipment on site	"	0.036		2.45	3.17	5.63
1200	Site grading (cut and fill to 6") less than 1 acre						
1300	75 hp dozer	CY	0.053		3.68	4.76	8.45
1400	1.5 c.y. backhoe/loader	"	0.080		5.52	7.15	12.75
31 - 23131	**BASE COURSE**						**31 - 23131**
1019	Base course, crushed stone						
1020	3" thick	SY	0.004	3.93	0.36	0.52	4.82
1030	4" thick	"	0.004	5.24	0.39	0.56	6.20
1040	6" thick	"	0.005	7.88	0.43	0.61	8.93
2500	Base course, bank run gravel						

EARTHWORK, EXCAVATION & FILL

ID Code	Component Descriptions	Unit of Meas.	Manhr / Unit	Material Cost	Labor Cost	Equipment Cost	Total Cost
	Description	**Output**		**Unit Costs**			
31 - 23131	**BASE COURSE, Cont'd...**						**31 - 23131**
3020	4" deep	SY	0.004	4.64	0.38	0.55	5.57
3040	6" deep	"	0.005	6.96	0.42	0.60	7.98
4000	Prepare and roll sub-base						
4020	Minimum	SY	0.004		0.36	0.52	0.89
4030	Average	"	0.005		0.45	0.65	1.11
4040	Maximum	"	0.007		0.61	0.87	1.48
31 - 23132	**BORROW**						**31 - 23132**
1000	Borrow fill, FOB at pit						
1005	Sand, haul to site, round trip						
1010	10 mile	CY	0.080	33.00	7.35	10.50	51.00
1020	20 mile	"	0.133	33.00	12.25	17.50	63.00
1030	30 mile	"	0.200	33.00	18.50	26.25	78.00
3980	Place borrow fill and compact						
4000	Less than 1 in 4 slope	CY	0.040	41.75	3.67	5.25	51.00
4100	Greater than 1 in 4 slope	"	0.053	41.75	4.90	7.00	54.00
31 - 23137	**GRAVEL AND STONE**						**31 - 23137**
0120	FOB at plant, material only						
1000	No. 21 crusher run stone	CY					49.50
1100	No. 26 crusher run stone	"					49.50
1140	No. 57 stone	"					49.50
1150	No. 67 gravel	"					49.50
1180	No. 68 stone	"					49.50
1220	No. 78 stone	"					49.50
1235	No. 78 gravel, (pea gravel)	"					49.50
1250	No. 357 or B-3 stone	"					49.50
1260	Structural & foundation backfill						
1400	No. 21 crusher run stone	TON					36.75
1500	No. 26 crusher run stone	"					36.75
1600	No. 57 stone	"					36.75
2160	No. 67 gravel	"					36.75
2210	No. 68 stone	"					36.75
2220	No. 78 stone	"					36.75
2280	No. 78 gravel (pea gravel)	"					36.75
3240	No. 357 or B-3 stone	"					36.75

EARTHWORK, EXCAVATION & FILL

		Output		Unit Costs			
	Description	Output		**Unit Costs**			
ID Code	Component Descriptions	Unit of Meas.	Manhr / Unit	Material Cost	Labor Cost	Equipment Cost	Total Cost
31 - 23163	**BULK EXCAVATION**						**31 - 23163**
1000	Excavation, by small dozer						
1020	Large areas	CY	0.016		1.10	1.43	2.53
1040	Small areas	"	0.027		1.84	2.38	4.22
1060	Trim banks	"	0.040		2.76	3.57	6.33
1700	Hydraulic excavator						
1720	1 c.y. capacity						
1740	Light material	CY	0.040		2.76	2.62	5.38
1760	Medium material	"	0.048		3.31	3.15	6.46
1780	Wet material	"	0.060		4.14	3.93	8.08
1790	Blasted rock	"	0.069		4.73	4.50	9.23
1800	1-1/2 c.y. capacity						
1820	Light material	CY	0.010		0.91	1.31	2.23
1840	Medium material	"	0.013		1.22	1.75	2.97
1860	Wet material	"	0.016		1.47	2.10	3.57
2000	Wheel mounted front-end loader						
2020	7/8 c.y. capacity						
2040	Light material	CY	0.020		1.83	2.62	4.46
2060	Medium material	"	0.023		2.10	3.00	5.10
2080	Wet material	"	0.027		2.45	3.50	5.95
2100	Blasted rock	"	0.032		2.94	4.20	7.14
2200	1-1/2 c.y. capacity						
2220	Light material	CY	0.011		1.05	1.50	2.55
2240	Medium material	"	0.012		1.13	1.61	2.74
2260	Wet material	"	0.013		1.22	1.75	2.97
2280	Blasted rock	"	0.015		1.33	1.90	3.24
2300	2-1/2 c.y. capacity						
2320	Light material	CY	0.009		0.86	1.23	2.10
2340	Medium material	"	0.010		0.91	1.31	2.23
2360	Wet material	"	0.011		0.98	1.40	2.38
2380	Blasted rock	"	0.011		1.05	1.50	2.55
2600	Track mounted front-end loader						
2620	1-1/2 c.y. capacity						
2640	Light material	CY	0.013		1.22	1.75	2.97
2660	Medium material	"	0.015		1.33	1.90	3.24
2680	Wet material	"	0.016		1.47	2.10	3.57
2700	Blasted rock	"	0.018		1.63	2.33	3.96
2720	2-3/4 c.y. capacity						
2740	Light material	CY	0.008		0.73	1.05	1.78

EARTHWORK, EXCAVATION & FILL

ID Code	Component Descriptions	Unit of Meas.	Manhr / Unit	Material Cost	Labor Cost	Equipment Cost	Total Cost
31 - 23163	**BULK EXCAVATION, Cont'd...**						**31 - 23163**
2760	Medium material	CY	0.009		0.81	1.16	1.98
2780	Wet material	"	0.010		0.91	1.31	2.23
2790	Blasted rock	"	0.011		1.05	1.50	2.55
31 - 23164	**BUILDING EXCAVATION**						**31 - 23164**
0090	Structural excavation, unclassified earth						
0100	3/8 c.y. backhoe	CY	0.107		9.80	14.00	23.75
0110	3/4 c.y. backhoe	"	0.080		7.35	10.50	17.75
0120	1 c.y. backhoe	"	0.067		6.12	8.75	14.75
0600	Foundation backfill and compaction by machine	"	0.160		14.75	21.00	35.75
31 - 23165	**HAND EXCAVATION**						**31 - 23165**
0980	Excavation						
1000	To 2' deep						
1020	Normal soil	CY	0.889		62.00		62.00
1040	Sand and gravel	"	0.800		56.00		56.00
1060	Medium clay	"	1.000		70.00		70.00
1080	Heavy clay	"	1.143		80.00		80.00
1100	Loose rock	"	1.333		93.00		93.00
1200	To 6' deep						
1220	Normal soil	CY	1.143		80.00		80.00
1240	Sand and gravel	"	1.000		70.00		70.00
1260	Medium clay	"	1.333		93.00		93.00
1280	Heavy clay	"	1.600		110		110
1300	Loose rock	"	2.000		140		140
2020	Backfilling foundation without compaction, 6" lifts	"	0.500		34.75		34.75
2200	Compaction of backfill around structures or in trench						
2220	By hand with air tamper	CY	0.571		39.75		39.75
2240	By hand with vibrating plate tamper	"	0.533		37.00		37.00
2250	1 ton roller	"	0.400		27.75	35.75	63.00
5400	Miscellaneous hand labor						
5440	Trim slopes, sides of excavation	SF	0.001		0.09		0.09
5450	Trim bottom of excavation	"	0.002		0.11		0.11
5460	Excavation around obstructions and services	CY	2.667		190		190

EARTHWORK, EXCAVATION & FILL

ID Code	Component Descriptions	Unit of Meas.	Manhr / Unit	Material Cost	Labor Cost	Equipment Cost	Total Cost
	Description	**Output**		**Unit Costs**			
31 - 23167	**UTILITY EXCAVATION**						**31 - 23167**
2080	Trencher, sandy clay, 8" wide trench						
2100	18" deep	LF	0.018		1.22	1.58	2.81
2200	24" deep	"	0.020		1.38	1.78	3.16
2300	36" deep	"	0.023		1.57	2.04	3.62
6080	Trench backfill, 95% compaction						
7000	Tamp by hand	CY	0.500		34.75		34.75
7050	Vibratory compaction	"	0.400		27.75		27.75
7060	Trench backfilling, with borrow sand, place & compact	"	0.400	33.00	27.75		61.00
31 - 23168	**ROADWAY EXCAVATION**						**31 - 23168**
0100	Roadway excavation						
0110	1/4 mile haul	CY	0.016		1.47	2.10	3.57
0120	2 mile haul	"	0.027		2.45	3.50	5.95
0130	5 mile haul	"	0.040		3.67	5.25	8.92
3000	Spread base course	"	0.020		1.83	2.62	4.46
3100	Roll and compact	"	0.027		2.45	3.50	5.95
31 - 23169	**HAULING MATERIAL**						**31 - 23169**
0090	Haul material by 10 c.y. dump truck, round trip distance						
0100	1 mile	CY	0.044		3.07	3.97	7.04
0110	2 mile	"	0.053		3.68	4.76	8.45
0120	5 mile	"	0.073		5.02	6.50	11.50
0130	10 mile	"	0.080		5.52	7.15	12.75
0140	20 mile	"	0.089		6.14	7.94	14.00
0150	30 mile	"	0.107		7.37	9.53	17.00
31 - 23336	**TRENCHING**						**31 - 23336**
0100	Trenching and continuous footing excavation						
0980	By gradall						
1000	1 c.y. capacity						
1020	Light soil	CY	0.023		2.10	3.00	5.10
1040	Medium soil	"	0.025		2.26	3.23	5.49
1060	Heavy/wet soil	"	0.027		2.45	3.50	5.95
1080	Loose rock	"	0.029		2.67	3.81	6.49
1090	Blasted rock	"	0.031		2.82	4.03	6.86
1095	By hydraulic excavator						
1100	1/2 c.y. capacity						
1120	Light soil	CY	0.027		2.45	3.50	5.95
1140	Medium soil	"	0.029		2.67	3.81	6.49

EARTHWORK, EXCAVATION & FILL

ID Code	Component Descriptions	Unit of Meas.	Manhr / Unit	Material Cost	Labor Cost	Equipment Cost	Total Cost
31 - 23336	**TRENCHING, Cont'd...**						**31 - 23336**
1160	Heavy/wet soil	CY	0.032		2.94	4.20	7.14
1180	Loose rock	"	0.036		3.26	4.66	7.93
1190	Blasted rock	"	0.040		3.67	5.25	8.92
1200	1 c.y. capacity						
1220	Light soil	CY	0.019		1.72	2.47	4.20
1240	Medium soil	"	0.020		1.83	2.62	4.46
1260	Heavy/wet soil	"	0.021		1.96	2.80	4.76
1280	Loose rock	"	0.023		2.10	3.00	5.10
1300	Blasted rock	"	0.025		2.26	3.23	5.49
1400	1-1/2 c.y. capacity						
1420	Light soil	CY	0.017		1.54	2.21	3.75
1440	Medium soil	"	0.018		1.63	2.33	3.96
1460	Heavy/wet soil	"	0.019		1.72	2.47	4.20
1480	Loose rock	"	0.020		1.83	2.62	4.46
1500	Blasted rock	"	0.021		1.96	2.80	4.76
1600	2 c.y. capacity						
1620	Light soil	CY	0.016		1.47	2.10	3.57
1640	Medium soil	"	0.017		1.54	2.21	3.75
1660	Heavy/wet soil	"	0.018		1.63	2.33	3.96
1680	Loose rock	"	0.019		1.72	2.47	4.20
1690	Blasted rock	"	0.020		1.83	2.62	4.46
3000	Hand excavation						
3100	Bulk, wheeled 100'						
3120	Normal soil	CY	0.889		62.00		62.00
3140	Sand or gravel	"	0.800		56.00		56.00
3160	Medium clay	"	1.143		80.00		80.00
3180	Heavy clay	"	1.600		110		110
3200	Loose rock	"	2.000		140		140
3300	Trenches, up to 2' deep						
3320	Normal soil	CY	1.000		70.00		70.00
3340	Sand or gravel	"	0.889		62.00		62.00
3360	Medium clay	"	1.333		93.00		93.00
3380	Heavy clay	"	2.000		140		140
3390	Loose rock	"	2.667		190		190
3400	Trenches, to 6' deep						
3420	Normal soil	CY	1.143		80.00		80.00
3440	Sand or gravel	"	1.000		70.00		70.00
3460	Medium clay	"	1.600		110		110

EARTHWORK, EXCAVATION & FILL

ID Code	Component Descriptions	Unit of Meas.	Manhr / Unit	Material Cost	Labor Cost	Equipment Cost	Total Cost
	Description	**Output**		**Unit Costs**			
31 - 23336	**TRENCHING, Cont'd...**						**31 - 23336**
3480	Heavy clay	CY	2.667		190		190
3500	Loose rock	"	4.000		280		280
3590	Backfill trenches						
3600	With compaction						
3620	By hand	CY	0.667		46.50		46.50
3640	By 60 hp tracked dozer	"	0.020		1.38	1.78	3.16

SOIL STABILIZATION & TREATMENT

ID Code	Component Descriptions	Unit of Meas.	Manhr / Unit	Material Cost	Labor Cost	Equipment Cost	Total Cost
31 - 31160	**SOIL TREATMENT**						**31 - 31160**
1100	Soil treatment, termite control pretreatment						
1120	Under slabs	SF	0.004	0.77	0.30		1.07
1140	By walls	"	0.005	0.77	0.37		1.14
31 - 32005	**SOIL STABILIZATION**						**31 - 32005**
0100	Straw bale secured with rebar	LF	0.027	8.31	1.85		10.25
0120	Filter barrier, 18" high filter fabric	"	0.080	2.01	5.56		7.57
0130	Sediment fence, 36" fabric with 6" mesh	"	0.100	4.76	6.96		11.75
1000	Soil stabilization with tar paper, burlap, straw and	SF	0.001	0.40	0.07		0.47
31 - 37001	**RIPRAP**						**31 - 37001**
0100	Riprap						
0110	Crushed stone blanket, max size 2-1/2"	TON	0.533	44.25	37.00	58.00	140
0120	Stone, quarry run, 300 lb. stones	"	0.492	55.00	34.25	54.00	140
0130	400 lb. stones	"	0.457	58.00	31.75	50.00	140
0140	500 lb. stones	"	0.427	60.00	29.75	46.75	140
0150	750 lb. stones	"	0.400	62.00	27.75	43.75	130
0160	Dry concrete riprap in bags 3" thick, 80 lb. per bag	BAG	0.027	7.45	1.85	2.91	12.25

PILES AND CAISSONS

ID Code	Component Descriptions	Unit of Meas.	Manhr / Unit	Material Cost	Labor Cost	Equipment Cost	Total Cost
31 - 62165	**STEEL PILES**						**31 - 62165**
1000	H-section piles						
1010	8x8						
1020	36 lb/ft						
1021	30' long	LF	0.080	32.75	7.10	9.16	49.00
1022	40' long	"	0.064	32.75	5.68	7.33	45.75
5000	Tapered friction piles, fluted casing, up to 50'						
5002	With 4000 psi concrete no reinforcing						

PILES AND CAISSONS

ID Code	Description — Component Descriptions	Output — Unit of Meas.	Manhr / Unit	Unit Costs — Material Cost	Labor Cost	Equipment Cost	Total Cost
31 - 62165	**STEEL PILES, Cont'd...**						**31 - 62165**
5040	12" dia.	LF	0.048	21.50	4.26	5.50	31.25
5060	14" dia.	"	0.049	24.75	4.37	5.64	34.75
31 - 62166	**STEEL PIPE PILES**						**31 - 62166**
1000	Concrete filled, 3000# concrete, up to 40'						
1100	8" dia.	LF	0.069	27.00	6.08	7.85	41.00
1120	10" dia.	"	0.071	34.75	6.31	8.14	49.25
1140	12" dia.	"	0.074	40.25	6.55	8.46	55.00
2000	Pipe piles, non-filled						
2020	8" dia.	LF	0.053	24.25	4.73	6.11	35.00
2040	10" dia.	"	0.055	30.75	4.87	6.28	42.00
2060	12" dia.	"	0.056	37.50	5.01	6.47	49.00
2520	Splice						
2540	8" dia.	EA	1.600	110	110		220
2560	10" dia.	"	1.600	120	110		230
2580	12" dia.	"	2.000	130	140		270
2680	Standard point						
2700	8" dia.	EA	1.600	140	110		250
2740	10" dia.	"	1.600	180	110		290
2760	12" dia.	"	2.000	190	140		330
2880	Heavy duty point						
2900	8" dia.	EA	2.000	230	140		370
2920	10" dia.	"	2.000	330	140		470
2940	12" dia.	"	2.667	350	190		540
31 - 62190	**WOOD AND TIMBER PILES**						**31 - 62190**
0080	Treated wood piles, 12" butt, 8" tip						
0100	25' long	LF	0.096	18.00	8.52	11.00	37.50
0110	30' long	"	0.080	19.25	7.10	9.16	35.50
0120	35' long	"	0.069	19.25	6.08	7.85	33.25
0125	40' long	"	0.060	19.25	5.32	6.87	31.50
31 - 63135	**PRESTRESSED PILING**						**31 - 63135**
0980	Prestressed concrete piling, less than 60' long						
1000	10" sq.	LF	0.040	20.75	3.55	4.58	29.00
1002	12" sq.	"	0.042	29.00	3.70	4.78	37.50
1480	Straight cylinder, less than 60' long						
1500	12" dia.	LF	0.044	27.00	3.87	5.00	35.75
1540	14" dia.	"	0.045	36.50	3.96	5.11	45.50

DIVISION 32
EXTERIOR
IMPROVEMENTS

PAVING

ID Code	Component Descriptions	Unit of Meas.	Manhr / Unit	Material Cost	Labor Cost	Equipment Cost	Total Cost
		Output		**Unit Costs**			

32 - 12000 ASPHALT SURFACES 32 - 12000

ID Code	Component Descriptions	Unit of Meas.	Manhr / Unit	Material Cost	Labor Cost	Equipment Cost	Total Cost
0050	Asphalt wearing surface, flexible pavement						
0100	1" thick	SY	0.016	4.97	1.42	1.83	8.22
0120	1-1/2" thick	"	0.019	7.50	1.70	2.20	11.50
1000	Binder course						
1010	1-1/2" thick	SY	0.018	7.10	1.57	2.03	10.75
1030	2" thick	"	0.022	9.43	1.93	2.50	13.75
2000	Bituminous sidewalk, no base						
2020	2" thick	SY	0.028	10.75	1.95	1.85	14.50
2040	3" thick	"	0.030	16.25	2.07	1.96	20.25

RIGID PAVING

32 - 13130 CONCRETE PAVING 32 - 13130

ID Code	Component Descriptions	Unit of Meas.	Manhr / Unit	Material Cost	Labor Cost	Equipment Cost	Total Cost
1080	Concrete paving, reinforced, 5000 psi concrete						
2000	6" thick	SY	0.150	33.25	13.25	17.25	64.00
2005	7" thick	"	0.160	38.75	14.25	18.25	71.00
2010	8" thick	"	0.171	44.25	15.25	19.75	79.00

32 - 13131 SIDEWALKS, CONCRETE 32 - 13131

ID Code	Component Descriptions	Unit of Meas.	Manhr / Unit	Material Cost	Labor Cost	Equipment Cost	Total Cost
6000	Walks, with wire mesh, base not incl.						
6010	4" thick	SF	0.027	2.51	1.85		4.36
6020	5" thick	"	0.032	3.39	2.22		5.61
6030	6" thick	"	0.040	4.18	2.78		6.96

UNIT PAVING

32 - 14160 PAVERS, MASONRY 32 - 14160

ID Code	Component Descriptions	Unit of Meas.	Manhr / Unit	Material Cost	Labor Cost	Equipment Cost	Total Cost
4010	Brick walk laid on sand, sand joints						
4020	Laid flat (4.5 per sf)	SF	0.089	3.78	7.53		11.25
4040	Laid on edge (7.2 per sf)	"	0.133	6.05	11.25		17.25
4080	Precast concrete patio blocks						
4100	2" thick						
5010	Natural	SF	0.027	2.68	2.26		4.94
5020	Colors	"	0.027	3.74	2.26		6.00
5080	Exposed aggregates, local aggregate						
5100	Natural	SF	0.027	8.08	2.26		10.25
5120	Colors	"	0.027	8.08	2.26		10.25
5130	Granite or limestone aggregate	"	0.027	8.44	2.26		10.75
5140	White tumblestone aggregate	"	0.027	6.05	2.26		8.31

UNIT PAVING

ID Code	Description		Output		Unit Costs			
	Component Descriptions	Unit of Meas.	Manhr / Unit	Material Cost	Labor Cost	Equipment Cost	Total Cost
32 - 14160	**PAVERS, MASONRY, Cont'd...**						**32 - 14160**
5960	Stone pavers, set in mortar						
5990	Bluestone						
6000	1" thick						
6010	Irregular	SF	0.200	6.14	17.00		23.25
6020	Snapped rectangular	"	0.160	9.35	13.50		22.75
6060	1-1/2" thick, random rectangular	"	0.200	10.75	17.00		27.75
6070	2" thick, random rectangular	"	0.229	12.75	19.25		32.00
6080	Slate						
6090	Natural cleft						
6100	Irregular, 3/4" thick	SF	0.229	7.71	19.25		27.00
6110	Random rectangular						
6120	1-1/4" thick	SF	0.200	16.75	17.00		33.75
6130	1-1/2" thick	"	0.222	18.75	18.75		37.50
6140	Granite blocks						
6150	3" thick, 3" to 6" wide						
7020	4" to 12" long	SF	0.267	11.00	22.50		33.50
7030	6" to 15" long	"	0.229	7.15	19.25		26.50
9800	Crushed stone, white marble, 3" thick	"	0.016	1.70	1.11		2.81

SITE IMPROVEMENTS

ID Code	**32 - 31130**	**CHAIN LINK FENCE**					**32 - 31130**
0230	Chain link fence, 9 ga., galvanized, with posts 10' o.c.						
0250	4' high	LF	0.057	7.75	3.97		11.75
0260	5' high	"	0.073	10.25	5.06		15.25
0270	6' high	"	0.100	11.75	6.96		18.75
1070	Corner or gate post, 3" post						
1080	4' high	EA	0.267	90.00	18.50		110
1084	5' high	"	0.296	100	20.50		120
1085	6' high	"	0.348	110	24.25		130
1100	Gate with gate posts, galvanized, 3' wide						
1102	4' high	EA	2.000	130	140		270
1104	5' high	"	2.667	170	190		360
1106	6' high	"	2.667	200	190		390
1161	Fabric, galvanized chain link, 2" mesh, 9 ga.						
1163	4' high	LF	0.027	5.05	1.85		6.90
1164	5' high	"	0.032	6.19	2.22		8.41
1165	6' high	"	0.040	8.66	2.78		11.50

SITE IMPROVEMENTS

ID Code	Description — Component Descriptions	Output — Unit of Meas.	Output — Manhr / Unit	Unit Costs — Material Cost	Unit Costs — Labor Cost	Unit Costs — Equipment Cost	Unit Costs — Total Cost
32 - 31130	**CHAIN LINK FENCE, Cont'd...**						**32 - 31130**
1400	Line post, no rail fitting, galvanized, 2-1/2" dia.						
1410	4' high	EA	0.229	29.25	16.00		45.25
1420	5' high	"	0.250	32.00	17.50		49.50
1430	6' high	"	0.267	35.00	18.50		54.00
1980	Vinyl coated, 9 ga., with posts 10' o.c.						
2000	4' high	LF	0.057	12.50	3.97		16.50
2010	5' high	"	0.073	15.00	5.06		20.00
2020	6' high	"	0.100	17.75	6.96		24.75
2160	Gate, with posts, 3' wide						
2180	4' high	EA	2.000	150	140		290
2190	5' high	"	2.667	170	190		360
2200	6' high	"	2.667	200	190		390
3000	Fabric, vinyl, chain link, 2" mesh, 9 ga.						
3010	4' high	LF	0.027	4.74	1.85		6.59
3020	5' high	"	0.032	5.79	2.22		8.01
3030	6' high	"	0.040	8.10	2.78		11.00
8000	Swing gates, galvanized, 4' high						
8010	Single gate						
8020	3' wide	EA	2.000	190	140		330
8025	4' wide	"	2.000	210	140		350
8155	6' high						
8158	Single gate						
8160	3' wide	EA	2.667	250	190		440
8165	4' wide	"	2.667	280	190		470
32 - 31901	**SHRUB & TREE MAINTENANCE**						**32 - 31901**
1000	Moving shrubs on site						
1220	3' high	EA	0.800		56.00		56.00
1240	4' high	"	0.889		62.00		62.00
2000	Moving trees on site						
3060	6' high	EA	0.533		36.75	35.00	72.00
3080	8' high	"	0.600		41.50	39.25	81.00
3100	10' high	"	0.800		55.00	53.00	110
3110	Palm trees						
3140	10' high	EA	0.800		55.00	53.00	110
3144	40' high	"	4.800		330	310	650

SITE IMPROVEMENTS

ID Code	Description Component Descriptions	Output		Unit Costs			
		Unit of Meas.	Manhr / Unit	Material Cost	Labor Cost	Equipment Cost	Total Cost
32 - 31902	**FERTILIZING**						**32 - 31902**
0080	Fertilizing (23#/1000 sf)						
0100	By square yard	SY	0.002	0.04	0.13	0.05	0.23
0120	By acre	ACRE	10.000	200	700	260	1,160
2980	Liming (70#/1000 sf)						
3000	By square yard	SY	0.003	0.04	0.18	0.07	0.29
3020	By acre	ACRE	13.333	220	930	350	1,500
32 - 31903	**WEED CONTROL**						**32 - 31903**
1000	Weed control, bromicil, 15 lb./acre, wettable powder	ACRE	4.000	330	280		610
1100	Vegetation control, by application of plant killer	SY	0.003	0.02	0.22		0.24
1200	Weed killer, lawns and fields	"	0.002	0.27	0.11		0.38
32 - 84004	**LAWN IRRIGATION**						**32 - 84004**
0480	Residential system, complete						
0490	Minimum	ACRE					24,060
0520	Maximum	"					45,790

PLANTING

ID Code	Description Component Descriptions	Output		Unit Costs			
		Unit of Meas.	Manhr / Unit	Material Cost	Labor Cost	Equipment Cost	Total Cost
32 - 91191	**TOPSOIL**						**32 - 91191**
0005	Spread topsoil, with equipment						
0010	Minimum	CY	0.080		7.35	10.50	17.75
0020	Maximum	"	0.100		9.19	13.00	22.25
0080	By hand						
0100	Minimum	CY	0.800		56.00		56.00
0110	Maximum	"	1.000		70.00		70.00
0980	Area prep. seeding (grade, rake and clean)						
1000	Square yard	SY	0.006		0.44		0.44
1020	By acre	ACRE	32.000		2,230		2,230
2000	Remove topsoil and stockpile on site						
2020	4" deep	CY	0.067		6.12	8.75	14.75
2040	6" deep	"	0.062		5.65	8.07	13.75
2200	Spreading topsoil from stock pile						
2220	By loader	CY	0.073		6.68	9.54	16.25
2240	By hand	"	0.800		74.00	110	180
2260	Top dress by hand	SY	0.008		0.73	1.05	1.78
2280	Place imported top soil						
2300	By loader						
2320	4" deep	SY	0.008		0.73	1.05	1.78

PLANTING

	Description	Output		Unit Costs			
ID Code	Component Descriptions	Unit of Meas.	Manhr / Unit	Material Cost	Labor Cost	Equipment Cost	Total Cost
32 - 91191	**TOPSOIL, Cont'd...**						**32 - 91191**
2340	6" deep	SY	0.009		0.81	1.16	1.98
2360	By hand						
2370	4" deep	SY	0.089		6.18		6.18
2380	6" deep	"	0.100		6.96		6.96
5980	Plant bed preparation, 18" deep						
6000	With backhoe/loader	SY	0.020		1.83	2.62	4.46
6010	By hand	"	0.133		9.28		9.28
32 - 92190	**SEEDING**						**32 - 92190**
0980	Mechanical seeding, 175 lb/acre						
1000	By square yard	SY	0.002	0.25	0.11	0.04	0.40
1020	By acre	ACRE	8.000	980	560	210	1,750
2040	450 lb/acre						
2060	By square yard	SY	0.002	0.62	0.13	0.05	0.81
2080	By acre	ACRE	10.000	2,430	700	260	3,390
5980	Seeding by hand, 10 lb per 100 SY						
6000	By square yard	SY	0.003	0.69	0.18		0.87
6010	By acre	ACRE	13.333	2,710	930		3,640
8010	Reseed disturbed areas	SF	0.004	0.06	0.27		0.33
32 - 93230	**PLANTS**						**32 - 93230**
0100	Euonymus coloratus, 18" (Purple Wintercreeper)	EA	0.133	3.48	9.28		12.75
0150	Hedera Helix, 2-1/4" pot (English ivy)	"	0.133	1.45	9.28		10.75
0200	Liriope muscari, 2" clumps	"	0.080	6.07	5.56		11.75
0250	Santolina, 12"	"	0.080	6.95	5.56		12.50
0280	Vinca major or minor, 3" pot	"	0.080	1.13	5.56		6.69
0300	Cortaderia argentia, 2 gallon (Pampas Grass)	"	0.080	22.00	5.56		27.50
0350	Ophiopogan japonicus, 1 quart (4" pot)	"	0.080	6.07	5.56		11.75
0400	Ajuga reptans, 2-3/4" pot (carpet bugle)	"	0.080	1.13	5.56		6.69
0450	Pachysandra terminalis, 2-3/4" pot (Japanese Spurge)	"	0.080	1.53	5.56		7.09
32 - 93330	**SHRUBS**						**32 - 93330**
0100	Juniperus conferia litoralis, 18"-24" (Shore Juniper)	EA	0.320	48.50	22.25		71.00
0150	Horizontalis plumosa, 18"-24" (Andorra Juniper)	"	0.320	51.00	22.25		73.00
0200	Sabina tamar-iscfolia-tamarix juniper, 18"-24"	"	0.320	51.00	22.25		73.00
0250	Chin San Jose, 18"-24" (San Jose Juniper)	"	0.320	51.00	22.25		73.00
0300	Sargenti, 18"-24" (Sargent's Juniper)	"	0.320	48.50	22.25		71.00
0350	Nandina domestica, 18"-24" (Heavenly Bamboo)	"	0.320	32.50	22.25		55.00
0400	Raphiolepis Indica Springtime, 18"-24"	"	0.320	35.00	22.25		57.00

PLANTING

ID Code	Component Descriptions	Unit of Meas.	Manhr / Unit	Material Cost	Labor Cost	Equipment Cost	Total Cost
	Description	**Output**		**Unit Costs**			
32 - 93330		**SHRUBS, Cont'd...**					**32 - 93330**
0450	Osmanthus Heterophyllus Gulftide, 18"-24"	EA	0.320	37.50	22.25		60.00
0460	Ilex Cornuta Burfordi Nana, 18"-24"	"	0.320	42.75	22.25		65.00
0550	Glabra, 18"-24" (Inkberry Holly)	"	0.320	40.25	22.25		63.00
0600	Azalea, Indica types, 18"-24"	"	0.320	45.25	22.25		68.00
0650	Kurume types, 18"-24"	"	0.320	51.00	22.25		73.00
0700	Berberis Julianae, 18"-24" (Wintergreen Barberry)	"	0.320	29.75	22.25		52.00
0800	Pieris Japonica Japanese, 18"-24"	"	0.320	29.75	22.25		52.00
0900	Ilex Cornuta Rotunda, 18"-24"	"	0.320	35.00	22.25		57.00
1000	Juniperus Horiz. Plumosa, 24"-30"	"	0.400	32.50	27.75		60.00
1200	Rhodopendrow Hybrids, 24"-30"	"	0.400	86.00	27.75		110
1400	Aucuba Japonica Varigata, 24"-30"	"	0.400	29.50	27.75		57.00
1600	Ilex Crenata Willow Leaf, 24"-30"	"	0.400	32.50	27.75		60.00
1620	Cleyera Japonica, 30"-36"	"	0.500	38.00	34.75		73.00
1700	Pittosporum Tobira, 30"-36"	"	0.500	44.25	34.75		79.00
1800	Prumus Laurocerasus, 30"-36"	"	0.500	82.00	34.75		120
1900	Ilex Cornuta Burfordi, 30"-36" (Burford Holly)	"	0.500	43.25	34.75		78.00
2000	Abelia Grandiflora, 24"-36" (Yew Podocarpus)	"	0.400	29.75	27.75		58.00
2100	Podocarpos Macrophylla, 24"-36"	"	0.400	48.50	27.75		76.00
2500	Pyracantha Coccinea Lalandi, 3'-4' (Firethorn)	"	0.500	27.75	34.75		63.00
2520	Photinia Frazieri, 3'-4' (Red Photinia)	"	0.500	44.00	34.75		79.00
2600	Forsythia Suspensa, 3'-4' (Weeping Forsythia)	"	0.500	27.75	34.75		63.00
2700	Camellia Japonica, 3'-4' (Common Camellia)	"	0.500	48.75	34.75		84.00
2800	Juniperus Chin Torulosa, 3'-4' (Hollywood Juniper)	"	0.500	52.00	34.75		87.00
2900	Cupressocyparis Leylandii, 3'-4'	"	0.500	43.75	34.75		79.00
3000	Ilex Opaca Fosteri, 5'-6' (Foster's Holly)	"	0.667	180	46.50		230
3200	Opaca, 5'-6' (American Holly)	"	0.667	250	46.50		300
3300	Nyrica Cerifera, 4'-5' (Southern Wax Myrtles)	"	0.571	55.00	39.75		95.00
3400	Ligustrum Japonicum, 4'-5' (Japanese Privet)	"	0.571	43.25	39.75		83.00
32 - 93430		**TREES**					**32 - 93430**
0100	Cornus Florida, 5'-6' (White flowering Dogwood)	EA	0.667	120	46.50		170
0120	Prunus Serrulata Kwanzan, 6'-8' (Kwanzan Cherry)	"	0.800	130	56.00		190
0130	Caroliniana, 6'-8' (Carolina Cherry Laurel)	"	0.800	150	56.00		210
0140	Cercis Canadensis, 6'-8' (Eastern Redbud)	"	0.800	100	56.00		160
0200	Koelreuteria Paniculata, 8'-10' (Goldenrain Tree)	"	1.000	180	70.00		250
0250	Acer Platanoides, 1-3/4"-2" (11'-13')	"	1.333	240	93.00		330
0300	Rubrum, 1-3/4"-2" (11'-13') (Red Maple)	"	1.333	180	93.00		270
0350	Saccharum, 1-3/4"-2" (Sugar Maple)	"	1.333	320	93.00		410

PLANTING

ID Code	Description Component Descriptions	Output Unit of Meas.	Manhr / Unit	Material Cost	Labor Cost	Equipment Cost	Total Cost
32 - 93430	**TREES, Cont'd...**						**32 - 93430**
0400	Fraxinus Pennsylvanica, 1-3/4"-2"	EA	1.333	150	93.00		240
0450	Celtis Occidentalis, 1-3/4"-2"	"	1.333	230	93.00		320
0460	Glenditsia Triacantos Inermis, 2"	"	1.333	210	93.00		300
1000	Prunus Cerasifera "Thundercloud", 6'-8'	"	0.800	120	56.00		180
1200	Yeodensis, 6'-8' (Yoshino Cherry)	"	0.800	130	56.00		190
1400	Lagerstroemia Indica, 8'-10' (Crapemyrtle)	"	1.000	210	70.00		280
1600	Crataegus Phaenopyrum, 8'-10'	"	1.000	320	70.00		390
1800	Quercus Borealis, 1-3/4"-2" (Northern Red Oak)	"	1.333	190	93.00		280
2000	Quercus Acutissima, 1-3/4"-2" (8'-10')	"	1.333	180	93.00		270
2100	Saliz Babylonica, 1-3/4"-2" (Weeping Willow)	"	1.333	90.00	93.00		180
2200	Tilia Cordata Greenspire, 1-3/4"-2" (10'-12')	"	1.333	400	93.00		490
2300	Malus, 2"-2-1/2" (8'-10') (Flowering Crabapple)	"	1.333	190	93.00		280
2400	Platanus Occidentalis, (12'-14')	"	1.600	300	110		410
2500	Pyrus Calleryana Bradford, 2"-2-1/2"	"	1.333	230	93.00		320
2600	Quercus Palustris, 2"-2-1/2" (12'-14') (Pin Oak)	"	1.333	260	93.00		350
2700	Phellos, 2-1/2"-3" (Willow Oak)	"	1.600	280	110		390
2800	Nigra, 2"-2-1/2" (Water Oak)	"	1.333	240	93.00		330
3000	Magnolia Soulangeana, 4'-5' (Saucer Magnolia)	"	0.667	140	46.50		190
3100	Grandiflora, 6'-8' (Southern Magnolia)	"	0.800	190	56.00		250
3200	Cedrus Deodara, 10'-12' (Deodare Cedar)	"	1.333	320	93.00		410
3300	Gingko Biloba, 10'-12' (2"-2-1/2")	"	1.333	300	93.00		390
3400	Pinus Thunbergi, 5'-6' (Japanese Black Pine)	"	0.667	120	46.50		170
3500	Strobus, 6'-8' (White Pine)	"	0.800	130	56.00		190
3600	Taeda, 6'-8' (Loblolly Pine)	"	0.800	110	56.00		170
3700	Quercus Virginiana, 2"-2-1/2" (Live Oak)	"	1.600	280	110		390
32 - 94002	**LANDSCAPE ACCESSORIES**						**32 - 94002**
0100	Steel edging, 3/16" x 4"	LF	0.010	5.33	0.69		6.02
0200	Landscaping stepping stones, 15"x15", white	EA	0.040	7.28	2.78		10.00
6000	Wood chip mulch	CY	0.533	51.00	37.00		88.00
6010	2" thick	SY	0.016	3.12	1.11		4.23
6020	4" thick	"	0.023	5.88	1.59		7.47
6030	6" thick	"	0.029	8.80	2.02		10.75
6200	Gravel mulch, 3/4" stone	CY	0.800	40.25	56.00		96.00
6300	White marble chips, 1" deep	SF	0.008	0.88	0.55		1.43
6980	Peat moss						
7000	2" thick	SY	0.018	4.76	1.23		5.99
7020	4" thick	"	0.027	9.16	1.85		11.00

PLANTING

ID Code	Component Descriptions	Unit of Meas.	Manhr / Unit	Material Cost	Labor Cost	Equipment Cost	Total Cost
	Description	**Output**		**Unit Costs**			
32 - 94002	**LANDSCAPE ACCESSORIES, Cont'd...**						**32 - 94002**
7030	6" thick	SY	0.033	14.00	2.32		16.25
7980	Landscaping timbers, treated lumber						
8000	4" x 4"	LF	0.027	4.48	1.85		6.33
8020	6" x 6"	"	0.029	10.50	1.98		12.50
8040	8" x 8"	"	0.033	12.75	2.32		15.00

DIVISION 33
UTILITIES

DISTRIBUTION PIPING

ID Code	Description / Component Descriptions	Output		Unit Costs			
		Unit of Meas.	Manhr / Unit	Material Cost	Labor Cost	Equipment Cost	Total Cost
33 - 11004	**DUCTILE IRON PIPE**					**33 - 11004**	
0990	Ductile iron pipe, cement lined, slip-on joints						
1000	4"	LF	0.067	21.25	4.60	4.37	30.25
1010	6"	"	0.071	24.50	4.87	4.63	34.00
1020	8"	"	0.075	32.00	5.18	4.92	42.00
1190	Mechanical joint pipe						
1200	4"	LF	0.092	44.50	6.37	6.05	57.00
1210	6"	"	0.100	53.00	6.91	6.56	67.00
1220	8"	"	0.109	70.00	7.53	7.15	85.00
1480	Fittings, mechanical joint						
1500	90 degree elbow						
1520	4"	EA	0.533	210	37.00		250
1540	6"	"	0.615	340	42.75		380
1560	8"	"	0.800	480	56.00		540
1700	45 degree elbow						
1720	4"	EA	0.533	220	37.00		260
1740	6"	"	0.615	300	42.75		340
1760	8"	"	0.800	420	56.00		480
33 - 11006	**PLASTIC PIPE**					**33 - 11006**	
0110	PVC, class 150 pipe						
0120	4" dia.	LF	0.060	8.30	4.14	3.93	16.50
0130	6" dia.	"	0.065	15.75	4.48	4.25	24.50
0140	8" dia.	"	0.069	25.00	4.73	4.50	34.25
0165	Schedule 40 pipe						
0170	1-1/2" dia.	LF	0.047	2.11	3.27		5.38
0180	2" dia.	"	0.050	2.68	3.48		6.16
0185	2-1/2" dia.	"	0.053	4.33	3.71		8.04
0190	3" dia.	"	0.057	5.51	3.97		9.48
0200	4" dia.	"	0.067	7.87	4.64		12.50
0210	6" dia.	"	0.080	14.25	5.56		19.75
0240	90 degree elbows						
0250	1"	EA	0.133	1.75	9.28		11.00
0260	1-1/2"	"	0.133	3.35	9.28		12.75
0270	2"	"	0.145	5.24	10.00		15.25
0280	2-1/2"	"	0.160	16.00	11.25		27.25
0290	3"	"	0.178	19.00	12.25		31.25
0300	4"	"	0.200	30.75	14.00		44.75
0310	6"	"	0.267	97.00	18.50		120

DISTRIBUTION PIPING

ID Code	Component Descriptions	Unit of Meas.	Manhr / Unit	Material Cost	Labor Cost	Equipment Cost	Total Cost
	Description		**Output**		**Unit Costs**		
33 - 11006	**PLASTIC PIPE, Cont'd...**						**33 - 11006**
0320	45 degree elbows						
0330	1"	EA	0.133	2.69	9.28		12.00
0340	1-1/2"	"	0.133	4.70	9.28		14.00
0350	2"	"	0.145	6.11	10.00		16.00
0360	2-1/2"	"	0.160	16.00	11.25		27.25
0370	3"	"	0.178	24.75	12.25		37.00
0380	4"	"	0.200	40.00	14.00		54.00
0390	6"	"	0.267	99.00	18.50		120
0400	Tees						
0410	1"	EA	0.160	2.32	11.25		13.50
0420	1-1/2"	"	0.160	4.46	11.25		15.75
0430	2"	"	0.178	6.44	12.25		18.75
0440	2-1/2"	"	0.200	21.25	14.00		35.25
0450	3"	"	0.229	28.00	16.00		44.00
0460	4"	"	0.267	45.50	18.50		64.00
0470	6"	"	0.320	150	22.25		170
0490	Couplings						
0510	1"	EA	0.133	1.42	9.28		10.75
0520	1-1/2"	"	0.133	2.04	9.28		11.25
0530	2"	"	0.145	3.14	10.00		13.25
0540	2-1/2"	"	0.160	6.90	11.25		18.25
0550	3"	"	0.178	10.75	12.25		23.00
0560	4"	"	0.200	14.00	14.00		28.00
0580	6"	"	0.267	44.50	18.50		63.00
1000	Drainage pipe						
1005	ABS						
1010	1-1/4" dia.	LF	0.047	2.03	3.27		5.30
1015	1-1/2" dia.	"	0.047	2.46	3.27		5.73
1020	2" dia.	"	0.050	3.14	3.48		6.62
1030	2-1/2" dia.	"	0.053	4.47	3.71		8.18
1040	3" dia.	"	0.057	5.26	3.97		9.23
1050	4" dia.	"	0.067	7.17	4.64		11.75
1055	6" dia.	"	0.080	12.00	5.56		17.50
1105	90 degree elbows						
1110	1-1/4"	EA	0.133	3.38	9.28		12.75
1125	1-1/2"	"	0.133	4.22	9.28		13.50
1135	2"	"	0.145	5.09	10.00		15.00
1145	2-1/2"	"	0.160	12.25	11.25		23.50

DISTRIBUTION PIPING

ID Code	Component Descriptions	Unit of Meas.	Manhr / Unit	Material Cost	Labor Cost	Equipment Cost	Total Cost
33 - 11006	**PLASTIC PIPE, Cont'd...**						**33 - 11006**
1155	3"	EA	0.178	12.50	12.25		24.75
1165	4"	"	0.200	22.25	14.00		36.25
1175	6"	"	0.267	48.75	18.50		67.00
1200	45 degree elbows						
1210	1-1/4"	EA	0.133	5.50	9.28		14.75
1220	1-1/2"	"	0.133	6.99	9.28		16.25
1230	2"	"	0.145	8.66	10.00		18.75
1240	2-1/2"	"	0.160	16.25	11.25		27.50
1250	3"	"	0.178	17.25	12.25		29.50
1260	4"	"	0.200	32.75	14.00		46.75
1270	6"	"	0.267	76.00	18.50		95.00

UTILITY SERVICES

ID Code	Component Descriptions	Unit of Meas.	Manhr / Unit	Material Cost	Labor Cost	Equipment Cost	Total Cost
33 - 21130	**WELLS**						**33 - 21130**
0980	Domestic water, drilled and cased						
1000	4" dia.	LF	0.480	32.00	42.50	55.00	130
1020	6" dia.	"	0.533	35.25	47.25	61.00	150

SANITARY SEWER

ID Code	Component Descriptions	Unit of Meas.	Manhr / Unit	Material Cost	Labor Cost	Equipment Cost	Total Cost
33 - 31003	**VITRIFIED CLAY PIPE**						**33 - 31003**
0100	Vitrified clay pipe, extra strength						
1020	6" dia.	LF	0.109	7.16	7.53	7.15	22.00
1040	8" dia.	"	0.114	8.59	7.89	7.50	24.00
1050	10" dia.	"	0.120	13.25	8.29	7.87	29.50
33 - 31004	**SANITARY SEWERS**						**33 - 31004**
0980	Clay						
1000	6" pipe	LF	0.080	9.41	5.52	5.25	20.25
2980	PVC						
3000	4" pipe	LF	0.060	4.05	4.14	3.93	12.25
3010	6" pipe	"	0.063	8.11	4.36	4.14	16.50

SANITARY SEWER

ID Code	Component Descriptions	Unit of Meas.	Manhr / Unit	Material Cost	Labor Cost	Equipment Cost	Total Cost
				Unit Costs			

ID Code	Component Descriptions	Unit of Meas.	Manhr / Unit	Material Cost	Labor Cost	Equipment Cost	Total Cost
33 - 36001	**DRAINAGE FIELDS**						**33 - 36001**
0080	Perforated PVC pipe, for drain field						
0100	4" pipe	LF	0.053	2.70	3.68	3.50	9.88
0120	6" pipe	"	0.057	5.08	3.94	3.75	12.75
33 - 36005	**SEPTIC TANKS**						**33 - 36005**
0980	Septic tank, precast concrete						
1000	1000 gals	EA	4.000	1,230	280	260	1,770
1200	2000 gals	"	6.000	3,320	410	390	4,130
1310	Leaching pit, precast concrete, 72" diameter						
1320	3' deep	EA	3.000	940	210	200	1,340
1340	6' deep	"	3.429	1,660	240	230	2,120
1360	8' deep	"	4.000	2,110	280	260	2,650
33 - 39133	**MANHOLES**						**33 - 39133**
0100	Precast sections, 48" dia.						
0110	Base section	EA	2.000	450	140	130	720
0120	1'0" riser	"	1.600	130	110	110	350
0130	1'4" riser	"	1.714	160	120	110	390
0140	2'8" riser	"	1.846	230	130	120	480
0150	4'0" riser	"	2.000	430	140	130	700
0160	2'8" cone top	"	2.400	280	170	160	600
0170	Precast manholes, 48" dia.						
0180	4' deep	EA	4.800	920	330	310	1,570
0200	6' deep	"	6.000	1,410	410	390	2,220
0250	7' deep	"	6.857	1,610	470	450	2,530
0260	8' deep	"	8.000	1,810	550	530	2,890
0280	10' deep	"	9.600	2,030	660	630	3,320
1000	Cast-in-place, 48" dia., with frame and cover						
1100	5' deep	EA	12.000	780	830	790	2,400
1120	6' deep	"	13.714	1,030	950	900	2,880
1140	8' deep	"	16.000	1,510	1,110	1,050	3,670
1160	10' deep	"	19.200	1,760	1,330	1,260	4,350
1480	Brick manholes, 48" dia. with cover, 8" thick						
1500	4' deep	EA	8.000	830	680		1,510
1501	6' deep	"	8.889	1,050	750		1,800
1505	8' deep	"	10.000	1,340	850		2,190
1510	10' deep	"	11.429	1,670	970		2,640
4200	Frames and covers, 24" diameter						
4210	300 lb	EA	0.800	470	56.00		530

SANITARY SEWER

ID Code	Description Component Descriptions	Output Unit of Meas.	Manhr / Unit	Material Cost	Labor Cost	Equipment Cost	Total Cost
33 - 39133	**MANHOLES, Cont'd...**						**33 - 39133**
4220	400 lb	EA	0.889	500	62.00		560
4980	Steps for manholes						
5000	7" x 9"	EA	0.160	20.00	11.25		31.25
5020	8" x 9"	"	0.178	25.25	12.25		37.50

STORMWATER MANAGEMENT

ID Code	Description Component Descriptions	Output Unit of Meas.	Manhr / Unit	Material Cost	Labor Cost	Equipment Cost	Total Cost
33 - 46190	**UNDERDRAIN**						**33 - 46190**
1480	Drain tile, clay						
1500	6" pipe	LF	0.053	5.19	3.68	3.50	12.25
1520	8" pipe	"	0.056	8.29	3.85	3.66	15.75
1580	Porous concrete, standard strength						
1600	6" pipe	LF	0.053	5.92	3.68	3.50	13.00
1620	8" pipe	"	0.056	6.40	3.85	3.66	14.00
1800	Corrugated metal pipe, perforated type						
1810	6" pipe	LF	0.060	8.62	4.14	3.93	16.75
1820	8" pipe	"	0.063	10.25	4.36	4.14	18.75
1980	Perforated clay pipe						
2000	6" pipe	LF	0.069	6.88	4.73	4.50	16.00
2020	8" pipe	"	0.071	8.01	4.87	4.63	17.50
2480	Drain tile, concrete						
2500	6" pipe	LF	0.053	4.70	3.68	3.50	12.00
2520	8" pipe	"	0.056	7.30	3.85	3.66	14.75
4980	Perforated rigid PVC underdrain pipe						
5000	4" pipe	LF	0.040	2.42	2.76	2.62	7.80
5100	6" pipe	"	0.048	4.64	3.31	3.15	11.00
5150	8" pipe	"	0.053	7.09	3.68	3.50	14.25
6980	Underslab drainage, crushed stone						
7000	3" thick	SF	0.008	0.36	0.55	0.52	1.43
7120	4" thick	"	0.009	0.49	0.63	0.60	1.73
7140	6" thick	"	0.010	0.75	0.69	0.65	2.09
7180	Plastic filter fabric for drain lines	"	0.008	0.55	0.55		1.10

HYDROCARBON STORAGE

ID Code	Description	Output		Unit Costs			
	Component Descriptions	Unit of Meas.	Manhr / Unit	Material Cost	Labor Cost	Equipment Cost	Total Cost
33 - 56001	**STORAGE TANKS**						**33 - 56001**
0080	Oil storage tank, underground, single wall, no excv.						
0090	Steel						
1000	500 gals	EA	3.000	5,260	210	200	5,660
1020	1,000 gals	"	4.000	7,150	280	260	7,690
1980	Fiberglass, double wall						
2000	550 gals	EA	4.000	14,870	280	260	15,410
2020	1,000 gals	"	4.000	19,120	280	260	19,660
2520	Above ground						
2530	Steel, single wall						
2540	275 gals	EA	2.400	2,990	170	160	3,310
2560	500 gals	"	4.000	7,480	280	260	8,020
2570	1,000 gals	"	4.800	10,200	330	310	10,850
3020	Fill cap	"	0.800	180	78.00		260
3040	Vent cap	"	0.800	180	78.00		260
3100	Level indicator	"	0.800	280	78.00		360

POWER & COMMUNICATIONS

ID Code	Description	Output		Unit Costs			
33 - 71160	**UTILITY POLES & FITTINGS**						**33 - 71160**
0980	Wood pole, creosoted						
1000	25'	EA	2.353	760	210		970
1010	30'	"	2.963	910	270		1,180
1065	Treated, wood preservative, 6"x6"						
1070	8'	EA	0.500	140	45.25		190
1080	10'	"	0.800	200	72.00		270
1090	12'	"	0.889	210	80.00		290
1100	14'	"	1.333	270	120		390
1120	16'	"	1.600	320	140		460
1140	18'	"	2.000	360	180		540
1150	20'	"	2.000	460	180		640
1155	Aluminum, brushed, no base						
1160	8'	EA	2.000	850	180		1,030
1170	10'	"	2.667	980	240		1,220
1180	15'	"	2.759	1,100	250		1,350
1190	20'	"	3.200	1,330	290		1,620
1200	25'	"	3.810	1,770	340		2,110
1235	Steel, no base						
1240	10'	EA	2.500	1,030	230		1,260

POWER & COMMUNICATIONS

ID Code	Component Descriptions	Unit of Meas.	Manhr / Unit	Material Cost	Labor Cost	Equipment Cost	Total Cost
33 - 71160	**UTILITY POLES & FITTINGS, Cont'd...**						**33 - 71160**
1245	15'	EA	2.963	1,130	270		1,400
1250	20'	"	3.810	1,510	340		1,850
1260	25'	"	4.520	1,700	410		2,110
2000	Concrete, no base						
2020	13'	EA	5.517	1,260	500		1,760
2040	16'	"	7.273	1,760	660		2,420
2060	18'	"	8.791	2,120	800		2,920
2080	25'	"	10.000	2,590	900		3,490

Description / *Output* / *Unit Costs* (column group headers)

DIVISION 48
ELECTRICAL POWER GENERATION

SOLAR

ID Code	Description — Component Descriptions	Output Unit of Meas.	Manhr / Unit	Unit Costs — Material Cost	Labor Cost	Equipment Cost	Total Cost
48 - 14110	**SOLAR ELECTRICAL SYSTEMS**						**48 - 14110**
1000	Photovoltaic, Full Grid-Tie System						
1010	Panel array, inverter, mounts, racks, conduit, etc.,						
1020	Minimum	EA					25,970
1030	Average	"					34,630
1040	Maximum	"					43,280
1050	4,000 Watt						
1060	Minimum	EA					34,630
1070	Average	"					46,170
1080	Maximum	"					57,710
1090	5,000 Watt						
1100	Minimum	EA					43,280
1110	Average	"					57,710
1120	Maximum	"					72,140
2000	Photovoltaic Components						
3000	Polycristalline Rigid Panel, 200 watt						
3010	Minimum	EA	0.286	580	25.75		610
3020	Average	"	0.286	650	25.75		680
3030	Maximum	"	0.286	330	25.75		360
3040	215 Watt						
3050	Minimum	EA	0.286	620	25.75		650
3060	Average	"	0.286	690	25.75		720
3070	Maximum	"	0.286	750	25.75		780
3080	230 Watt						
3090	Minimum	EA	0.286	680	25.75		710
3100	Average	"	0.286	750	25.75		780
3110	Maximum	"	0.286	820	25.75		850
3120	245 Watt						
3130	Minimum	EA	0.286	720	25.75		750
3140	Average	"	0.286	790	25.75		820
3150	Maximum	"	0.286	870	25.75		900
4000	Anodized aluminum rail, 8'						
4010	Minimum	EA	0.178	49.00	16.00		65.00
4020	Average	"	0.178	52.00	16.00		68.00
4030	Maximum	"	0.178	55.00	16.00		71.00
4040	10'						
4050	Minimum	EA	0.200	61.00	18.00		79.00
4060	Average	"	0.200	65.00	18.00		83.00
4070	Maximum	"	0.200	69.00	18.00		87.00

SOLAR

ID Code	Component Descriptions	Unit of Meas.	Manhr / Unit	Material Cost	Labor Cost	Equipment Cost	Total Cost
	Description	**Output**		**Unit Costs**			
48 - 14110	**SOLAR ELECTRICAL SYSTEMS, Cont'd...**						**48 - 14110**
4080	12'						
4090	Minimum	EA	0.229	75.00	20.75		96.00
4100	Average	"	0.229	78.00	20.75		99.00
4110	Maximum	"	0.229	81.00	20.75		100
4500	Panel mounts, aluminum, mount tile, flush						
4510	Minimum	EA	0.533	40.50	48.25		89.00
4520	Average	"	0.533	43.25	48.25		92.00
4530	Maximum	"	0.533	46.25	48.25		95.00
4540	Standard						
4550	Minimum	EA	0.533	31.75	48.25		80.00
4560	Average	"	0.533	34.50	48.25		83.00
4570	Maximum	"	0.533	37.50	48.25		86.00
4800	Rail clamp, mid-clamp						
4810	Minimum	EA					10.00
4820	Average	"					11.50
4830	Maximum	"					13.00
4840	End-clamp						
4850	Minimum	EA					13.00
4860	Average	"					14.75
4870	Maximum	"					16.25
4880	Anodized aluminum, rail splice kit						
4890	Minimum	EA					20.25
4900	Average	"					23.00
4910	Maximum	"					26.00
5000	Power Distribution Panel						
5010	Minimum	EA	2.000	170	180		350
5020	Average	"	2.000	210	180		390
5030	Maximum	"	2.000	240	180		420
6000	Inverters						
6010	Light capacity, micro inverter, 190 Watt						
6020	Minimum	EA	0.267	260	24.00		280
6030	Average	"	0.267	290	24.00		310
6040	Maximum	"	0.267	320	24.00		340
6500	Medium capacity, inverter, 1000 Watt						
6510	Minimum	EA	4.000	2,600	360		2,960
6520	Average	"	4.000	2,890	360		3,250
6530	Maximum	"	4.000	3,170	360		3,530
6540	2500 Watt						

SOLAR

ID Code	Description		Output		Unit Costs			
	Component Descriptions		Unit of Meas.	Manhr / Unit	Material Cost	Labor Cost	Equipment Cost	Total Cost

48 - 14110	**SOLAR ELECTRICAL SYSTEMS, Cont'd...**						**48 - 14110**
6550	Minimum	EA	4.000	4,330	360		4,690
6560	Average	"	4.000	4,620	360		4,980
6570	Maximum	"	4.000	4,910	360		5,270
6580	5000 Watt						
6590	Minimum	EA	4.000	6,930	360		7,290
6600	Average	"	4.000	7,210	360		7,570
6610	Maximum	"	4.000	7,500	360		7,860
7000	Circuits						
7010	Combiner Box, 12 circuit						
7020	Minimum	EA	2.667	790	240		1,030
7030	Average	"	2.667	950	240		1,190
7040	Maximum	"	2.667	1,050	240		1,290
7050	28 circuit						
7060	Minimum	EA	2.667	1,280	240		1,520
7070	Average	"	2.667	1,440	240		1,680
7080	Maximum	"	2.667	1,590	240		1,830
7200	Solar Circuit Breaker, 15A, 150 VDC						
7210	Minimum	EA					15.75
7220	Average	"					18.75
7230	Maximum	"					21.75
7240	Wires and Conductors						
8230	Cable, bare copper, 10 AWG, 500' coils						
8240	Minimum	EA					220
8250	Average	"					240
8260	Maximum	"					250
8270	Multi-Contact branch connector, MC4						
8280	Minimum	EA	0.267	40.50	24.00		65.00
8300	Average	"	0.267	43.25	24.00		67.00
8320	Maximum	"	0.267	46.25	24.00		70.00

Metro Area Multipliers

The costs as presented in this book attempt to represent national averages. Costs, however, vary among regions, states and even between adjacent localities.

In order to more closely approximate the probable costs for specific locations throughout the U.S., this table of Metro Area Multipliers is provided. These adjustment factors can be used to modify costs obtained from this book to help account for regional variations of construction costs and to provide a more accurate estimate for specific areas. The factors are formulated by comparing costs in a specific area to the costs presented in this Costbook. An example of how to use these factors is shown below. Whenever local current costs are known, whether material prices or labor rates, they should be used when more accuracy is required.

Cost from Costbook Pages   **X**   **Metro Area Multiplier**   **=**   **Adjusted Cost**

For example, a project estimated to cost $1,000,000 using the Costbook pages can be adjusted to more closely approximate the cost in Los Angeles:

$1,000,000   X   1.20   = $1,200,000

State	Metropolitan Area	Multiplier
AK	ANCHORAGE	1.09
	FAIRBANKS	1.09
	JUNEAU	1.09
	KETCHIKAN	1.09
	KODIAK	1.09
	NOME	1.09
	OTHER	1.09
	SITKA	1.09
AL	ANNISTON	0.83
	AUBURN	0.83
	BIRMINGHAM	0.84
	DECATUR	0.84
	DOTHAN	0.81
	FLORENCE	0.84
	GASDEN	0.83
	HUNTSVILLE	0.82
	MOBILE	0.83
	MONTGOMERY	0.83
	OPELIKA	0.83
	OTHER	0.83
	PRATTVILLE	0.83
	SELMA	0.83
	TUSCALOOSA	0.84
AR	BATESVILLE	0.80
	CONWAY	0.80
	EL DORADO	0.80
	FAYETTEVILLE	0.80
	FORT SMITH	0.80
	HOT SPRINGS	0.80
	JONESBORO	0.79
	LITTLE ROCK	0.82
	NORTH LITTLE ROCK	0.82
	OTHER	0.80
	PINE BLUFF	0.80
	ROGERS	0.78
	SPRINGDALE	0.80
	TEXARKANA	0.80

State	Metropolitan Area	Multiplier
AZ	CASA GRANDE	0.83
	CLIFTON	0.83
	FLAGSTAFF	0.84
	LAKE HAVASU CITY	0.83
	MESA	0.85
	OTHER	0.83
	PHOENIX	0.85
	SIERRA VISTA	0.83
	TUCSON	0.82
	YUMA	0.81
CA	ANAHEIM	1.21
	BAKERSFIELD	1.20
	CHICO	1.17
	EUREKA	1.17
	FAIRFIELD	1.17
	FRESNO	1.21
	LODI	1.21
	LONG BEACH	1.20
	LOS ANGELES	1.20
	MERCED	1.21
	MODESTO	1.21
	NAPA	1.17
	OAKLAND	1.21
	ORANGE COUNTY	1.21
	OTHER	1.20
	PARADISE	1.17
	PORTERVILLE	1.07
	REDDING	1.17
	RIVERSIDE	1.20
	SACRAMENTO	1.17
	SALINAS	1.21
	SAN BERNARDINO	1.21
	SAN DIEGO	1.18
	SAN FRANCISCO	1.21
	SAN JOSE	1.21
	SAN LUIS OBISPO	1.20
	SANTA BARBARA	1.20
	SANTA CRUZ	1.21
	SANTA ROSA	1.17

State	Metropolitan Area	Multiplier
CA	STOCKTON	1.21
	TULARE	1.07
	VALLEJO-FAIRFIELD-NAPA	1.17
	VENTURA	1.20
	VISALIA	1.07
	WATSONVILLE	1.21
	YOLO	1.17
	YUBA CITY	1.17
CO	BOULDER	0.91
	COLORADO SPRINGS	0.90
	DENVER	0.90
	DURANGO	0.90
	FORT COLLINS	0.88
	GRAND JUNCTION	0.88
	GREELEY	0.89
	LONGMONT	0.91
	LOVELAND	0.88
	OTHER	0.90
	PUEBLO	0.89
	STERLING	0.90
CT	BRIDGEPORT	1.11
	DANBURY	1.11
	HARTFORD	1.11
	MANCHESTER	1.11
	MERIDEN	1.10
	MIDDLETOWN	1.11
	NEW HAVEN	1.10
	NEW LONDON	1.10
	NORWALK	1.11
	NORWICH	1.10
	OTHER	1.10
	STAMFORD	1.11
	TORRINGTON	1.10
	WATERBURY	1.10
DC	WASHINGTON DC	1.00
DE	DOVER	1.08
	NEWARK	1.07
	OTHER	1.07
	WILMINGTON	1.07

State	Metropolitan Area	Multiplier
FL	BLOUNTSTOWN	0.86
	BOCA RATON	0.86
	BOSTWICK	0.90
	BRADENTON	0.87
	CAPE CORAL	0.86
	CLEARWATER	0.86
	DAYTONA BEACH	0.85
	FORT LAUDERDALE	0.88
	FORT MYERS	0.86
	FORT PIERCE	0.86
	FORT WALTON BEACH	0.85
	GAINESVILLE	0.85
	JACKSONVILLE	0.87
	LAKELAND	0.84
	MELBOURNE	0.91
	MIAMI	0.87
	NAPLES	0.86
	OCALA	0.85
	ORLANDO	0.84
	OTHER	0.86
	PALM BAY	0.91
	PANAMA CITY	0.86
	PENSACOLA	0.84
	PORT ST. LUCIE	0.86
	PUNTA GORDA	0.86
	SARASOTA	0.86
	ST PETERSBURG	0.86
	TALLAHASSEE	0.84
	TAMPA	0.85
	TITUSVILLE	0.91
	WEST PALM BEACH	0.86
	WINTER HAVEN	0.84
GA	ALBANY	0.86
	ATHENS	0.84
	ATLANTA	0.84
	AUGUSTA	0.84
	COLUMBUS	0.82
	MACON	0.84
	MARIETTA	0.84

State	Metropolitan Area	Multiplier
GA	OTHER	0.84
	ROME	0.84
	SAVANNAH	0.85
	VALDOSTA	0.84
HI	HILO	1.16
	HONOLULU	1.16
	MAUI	1.16
	OTHER	1.16
IA	BURLINGTON	0.93
	CEDAR FALLS	0.89
	CEDAR RAPIDS	0.92
	COUNCIL BLUFFS	0.93
	DAVENPORT	0.96
	DES MOINES	0.96
	DUBUQUE	0.94
	FT. DODGE	0.93
	IOWA CITY	0.93
	MARSHALLTOWN	0.93
	MASON CITY	0.93
	OTHER	0.93
	OTTUMWA	0.93
	SIOUX CITY	0.92
	WATERLOO	0.89
ID	BOISE CITY	0.87
	LEWISTON	0.88
	OTHER	0.88
	POCATELLO	0.89
	SALMON	0.88
	ST. ANTHONY	0.88
	ST. MARIES	0.88
	TWIN FALLS	0.88
IL	ALTON	1.05
	BLOOMINGTON	1.04
	CHAMPAIGN	1.02
	CHICAGO	1.19
	DE KALB	1.04
	DECATUR	1.00
	EAST ST. LOUIS	1.05
	FREEPORT	1.04
	HARRISBURG	1.02

State	Metropolitan Area	Multiplier
IL	KANKAKEE	1.18
	MOLINE	1.04
	NORMAL	1.04
	OTHER	1.05
	PEKIN	1.01
	PEORIA	1.01
	ROCKFORD	1.04
	SPRINGFIELD	1.00
	URBANA	1.02
IN	BLOOMINGTON	0.94
	COLUMBUS	0.94
	ELKHART	1.00
	EVANSVILLE	0.95
	FORT WAYNE	0.94
	GARY	1.02
	GOSHEN	1.00
	INDIANAPOLIS	0.94
	KOKOMO	0.94
	LAFAYETTE	0.94
	MUNCIE	0.94
	NEW ALBANY	0.97
	OTHER	0.97
	SOUTH BEND	1.02
	TERRE HAUTE	0.95
KS	DODGE CITY	1.01
	HUTCHINSON	1.01
	KANSAS CITY	1.04
	LAWRENCE	1.02
	MANHATTAN	1.01
	OLATHE	1.01
	OTHER	1.01
	SALINA	1.01
	TOPEKA	1.00
	WICHITA	0.95
KY	BOWLING GREEN	0.92
	HOPKINSVILLE	0.92
	LEXINGTON	0.93
	LOUISVILLE	0.91
	MAYFIELD	0.92
	MAYSVILLE	0.92

State	Metropolitan Area	Multiplier
KY	OTHER	0.92
	OWENSBORO	0.92
	PADUCAH	0.92
LA	ALEXANDRIA	0.84
	BATON ROUGE	0.85
	BOSSIER CITY	0.84
	HOUMA	0.85
	LAFAYETTE	0.84
	LAKE CHARLES	0.84
	MONROE	0.83
	NEW IBERIA	0.84
	NEW ORLEANS	0.85
	OTHER	0.84
	SHREVEPORT	0.85
MA	ATTLEBORO	1.17
	BARNSTABLE	1.17
	BOSTON	1.17
	BROCKTON	1.21
	FALL RIVER	1.17
	FITCHBURG	1.18
	FRAMINGHAM	1.17
	HAVERHILL	1.17
	LAWRENCE	1.17
	LEOMINSTER	1.18
	LOWELL	1.17
	NEW BEDFORD	1.17
	NORTHAMPTON	1.06
	OTHER	1.14
	PITTSFIELD	1.06
	SPRINGFIELD	1.06
	WORCESTER	1.18
	YARMOUTH	1.17
MD	ABERDEEN	0.89
	ANNAPOLIS	0.89
	BALTIMORE	0.91
	CUMBERLAND	0.87
	FREDERICK	0.89
	HAGERSTOWN	0.89
	LEXINGTON PARK	0.89
	OTHER	0.89

State	Metropolitan Area	Multiplier
MD	ROCKVILLE	0.89
	SALISBURY	0.89
ME	AUBURN	0.84
	AUGUSTA	0.83
	BANGOR	0.82
	BIDDEFORD	0.83
	FARMINGTON	0.83
	LEWISTOWN	0.84
	OTHER	0.83
	PORTLAND	0.84
	SANFORD	0.83
MI	ANN ARBOR	1.03
	BATTLE CREEK	0.94
	BAY CITY	0.93
	BENTON HARBOR	0.95
	CHEBOYGAN	0.94
	DETROIT	1.02
	EAST LANSING	0.96
	FLINT	0.98
	GRAND RAPIDS	0.84
	HOLLAND	0.88
	JACKSON	0.97
	KALAMAZOO	0.94
	LANSING	0.96
	MARQUETTE	0.94
	MIDLAND	0.92
	MUSKEGON	0.87
	OTHER	0.94
	SAGINAW	0.92
MN	AUSTIN	1.07
	DULUTH	1.05
	GRAND FORKS	1.07
	GRAND RAPIDS	1.07
	MANKATO	1.07
	MINNEAPOLIS	1.09
	MOORHEAD	1.07
	OTHER	1.07
	ROCHESTER	1.05
	ST CLOUD	1.05
	ST PAUL	1.09

State	Metropolitan Area	Multiplier
MO	COLUMBIA	1.00
	FARMINGTON	1.00
	JEFFERSON CITY	1.00
	JOPLIN	0.91
	KANSAS CITY	1.06
	OTHER	1.00
	POPLAR BLUFF	1.00
	SAINT JOSEPH	1.05
	SAINT LOUIS	1.02
	SPRINGFIELD	0.90
MS	BILOXI	0.82
	COLUMBUS	0.82
	GREENVILLE	0.82
	GULFPORT	0.82
	HATTIESBURG	0.83
	HELENA	0.84
	JACKSON	0.80
	MERIDIAN	0.82
	OTHER	0.82
	PASCAGOULA	0.84
	VICKSBURG	0.82
MT	BILLINGS	0.91
	BUTTE	0.92
	GLASGOW	0.92
	GREAT FALLS	0.92
	HAVRE	0.92
	HELENA	0.92
	KALISPELL	0.92
	MILES CITY	0.92
	MISSOULA	0.92
	OTHER	0.92
NC	ASHEVILLE	0.79
	CHAPEL HILL	0.78
	CHARLOTTE	0.79
	DURHAM	0.78
	FAYETTEVILLE	0.79
	GASTONIA	0.79
	GOLDSBORO	0.79
	GREENSBORO	0.78
	GREENVILLE	0.78

State	Metropolitan Area	Multiplier
NC	HICKORY	0.79
	HIGH POINT	0.78
	JACKSONVILLE	0.79
	KANNAPOLIS	0.79
	LENOIR	0.80
	MORGANTON	0.80
	OTHER	0.79
	RALEIGH	0.79
	ROCKY MOUNT	0.78
	WILMINGTON	0.78
	WINSTON-SALEM	0.78
ND	BISMARCK	0.89
	DICKINSON	0.91
	FARGO	0.92
	GRAND FORKS	0.90
	JAMESTOWN	0.91
	MINOT	0.91
	OTHER	0.91
	WILLISTON	0.91
NE	BEATRICE	0.91
	CHADRON	0.91
	COLUMBUS	0.91
	GRAND ISLAND	0.91
	LINCOLN	0.89
	NORFOLK	0.91
	NORTH PLATTE	0.91
	OMAHA	0.93
	OTHER	0.91
	SCOTTSBLUFF	0.91
	VALENTINE	0.91
NH	BERLIN	0.89
	CLAREMONT	0.89
	CONCORD	0.89
	KEENE	0.89
	LACONIA	0.89
	MANCHESTER	0.89
	NASHUA	0.89
	OTHER	0.89
	PORTSMOUTH	0.90
	ROCHESTER	0.89

State	Metropolitan Area	Multiplier
NJ	ATLANTIC CITY	1.20
	BERGEN	1.24
	BRIDGETON	1.19
	CAMDEN	1.21
	CAPE MAY	1.18
	FLANDERS	1.21
	HAMMONTON	1.20
	HUNTERDON	1.20
	JERSEY CITY	1.23
	LAKEWOOD	1.21
	MIDDLESEX	1.20
	MILLVILLE	1.19
	MONMOUTH	1.19
	NEW BRUNSWICK	1.21
	NEWARK	1.22
	OCEAN	1.18
	OTHER	1.21
	PASSAIC	1.23
	PATERSON	1.23
	SOMERSET	1.22
	TRENTON	1.20
	VINELAND	1.19
	WASHINGTON	1.21
	WILLINGBORO	1.21
NM	ALAMOGORDO	0.88
	ALBUQUERQUE	0.88
	CARLSBAD	0.88
	FARMINGTON	0.88
	GALLUP	0.88
	LAS CRUCES	0.87
	LAS VEGAS	0.88
	OTHER	0.88
	SANTA FE	0.88
NV	CARSON CITY	1.14
	ELY	1.14
	HAWTHORNE	1.14
	HENDERSON	1.15
	LAS VEGAS	1.15
	OTHER	1.14
	RENO	1.14

State	Metropolitan Area	Multiplier
NV	SPARKS	1.14
	WINNEMUCCA	1.14
NY	ALBANY	1.06
	AMSTERDAM	1.06
	BINGHAMTON	1.06
	BUFFALO	1.04
	DUTCHESS COUNTY	1.18
	ELMIRA	1.04
	GLENS FALLS	1.08
	ITHACA	1.11
	JAMESTOWN	1.11
	KINGSTON	1.18
	LOCKPORT	0.97
	LONG ISLAND	1.27
	MALONE	1.11
	NASSAU	1.27
	NEW YORK CITY	1.38
	NEWBURGH	1.18
	NIAGARA FALLS	0.97
	OTHER	1.11
	ROCHESTER	1.04
	ROME	0.94
	SCHENECTADY	1.06
	SUFFOLK	1.27
	SYRACUSE	0.96
	TROY	1.06
	UTICA	0.94
	WATERTOWN	1.11
	WHITE PLAINS	1.11
OH	AKRON	0.97
	CANTON	0.94
	CINCINNATI	0.94
	CLEVELAND	0.99
	COLUMBUS	0.95
	DAYTON	0.93
	ELYRIA	0.99
	FINDLAY	0.95
	HAMILTON	0.95
	LIMA	0.93
	LORAIN	0.99

State	Metropolitan Area	Multiplier
OH	MANSFIELD	0.95
	MARION	0.95
	MASSILLON	0.94
	MIDDLETOWN	0.95
	OTHER	0.95
	PORTSMOUTH	0.95
	SPRINGFIELD	0.94
	STEUBENVILLE	0.94
	TOLEDO	0.97
	WARREN	0.94
	YOUNGSTOWN	0.94
OK	BARTLESVILLE	0.87
	ENID	0.85
	LAWTON	0.86
	MUSKOGEE	0.87
	NORMAN	0.87
	OKLAHOMA CITY	0.87
	OTHER	0.87
	PONCA CITY	0.87
	TULSA	0.87
OR	ASHLAND	1.00
	BEND	1.05
	CORVALLIS	1.05
	EUGENE	1.05
	MEDFORD	1.00
	OTHER	1.05
	PENDLETON	1.05
	PORTLAND	1.08
	SALEM	1.06
	SPRINGFIELD	1.05
	THE DALLES	1.05
PA	ALLENTOWN	1.00
	ALTOONA	0.97
	BETHLEHEM	1.02
	CARLISLE	0.97
	EASTON	1.02
	ERIE	0.97
	HARRISBURG	0.98
	HAZLETON	0.99
	JOHNSTOWN	0.97

State	Metropolitan Area	Multiplier
PA	LANCASTER	0.97
	LEBANON	0.97
	NEW CASTLE	1.00
	OTHER	1.00
	PHILADELPHIA	1.14
	PITTSBURGH	1.02
	READING	1.02
	SCRANTON	0.99
	SHARON	0.99
	STATE COLLEGE	0.98
	WILKES-BARRE	0.99
	WILLIAMSPORT	0.99
	YORK	0.97
PR	MAYAGUEZ	0.73
	OTHER	0.73
	PONCE	0.73
	SAN JUAN	0.73
RI	NEWPORT	1.07
	OTHER	1.07
	PAWTUCKET	1.07
	PROVIDENCE	1.07
	WARWICK	1.07
	WESTERLY	1.07
	WOONSOCKET	1.07
SC	AIKEN	0.90
	ANDERSON	0.78
	CHARLESTON	0.79
	COLUMBIA	0.78
	FLORENCE	0.80
	GREENVILLE	0.78
	MYRTLE BEACH	0.77
	NORTH CHARLESTON	0.79
	OTHER	0.81
	ROCK HILL	0.81
	SPARTANBURG	0.79
	SUMTER	0.78
SD	ABERDEEN	0.84
	BROOKINGS	0.84
	MITCHELL	0.84
	OTHER	0.84

State	Metropolitan Area	Multiplier
SD	PIERRE	0.84
	RAPID CITY	0.84
	SIOUX FALLS	0.84
	WATERTOWN	0.84
TN	CHATTANOOGA	0.86
	CLARKSVILLE	0.86
	JACKSON	0.86
	JOHNSON CITY	0.84
	KNOXVILLE	0.87
	MEMPHIS	0.87
	NASHVILLE	0.85
	OTHER	0.86
TX	ABILENE	0.84
	AMARILLO	0.84
	ARLINGTON	0.82
	AUSTIN	0.86
	BEAUMONT	0.85
	BRAZORIA	0.84
	BROWNSVILLE	0.83
	BRYAN	0.87
	COLLEGE STATION	0.87
	CORPUS CHRISTI	0.84
	DALLAS	0.84
	DENISON	0.85
	EDINBURG	0.84
	EL PASO	0.83
	FORT WORTH	0.82
	GALVESTON	0.84
	HARLINGEN	0.83
	HOUSTON	0.86
	KILLEEN	0.81
	LAREDO	0.84
	LONGVIEW	0.84
	LUBBOCK	0.85
	MARSHALL	0.81
	MCALLEN	0.84
	MIDLAND	0.84
	MISSION	0.84
	ODESSA	0.84
	OTHER	0.84

State	Metropolitan Area	Multiplier
TX	PORT ARTHUR	0.85
	SAN ANGELO	0.83
	SAN ANTONIO	0.88
	SAN BENITO	0.83
	SAN MARCOS	0.85
	SHERMAN	0.85
	TEMPLE	0.81
	TEXARKANA	0.83
	TEXAS CITY	0.84
	TYLER	0.82
	VICTORIA	0.84
	WACO	0.84
	WICHITA FALLS	0.85
UT	LOGAN	0.86
	OGDEN	0.86
	OREM	0.85
	OTHER	0.86
	PROVO	0.85
	SALT LAKE CITY	0.87
VA	ALEXANDRIA	0.85
	ARLINGTON	0.85
	CHARLOTTESVILLE	0.85
	LYNCHBURG	0.84
	NEWPORT NEWS	0.86
	NORFOLK	0.85
	OTHER	0.85
	PETERSBURG	0.85
	RICHMOND	0.85
	ROANOKE	0.84
	VIRGINIA BEACH	0.85
VT	BARRE	0.83
	BRATTLEBORO	0.83
	BURLINGTON	0.83
	NEWPORT	0.83
	OTHER	0.83
	RUTLAND	0.83
	SPRINGFIELD	0.83
	ST. ALBANS	0.83
	ST. JOHNSBURY	0.83

State	Metropolitan Area	Multiplier
WA	BELLEVUE	1.15
	BELLINGHAM	1.09
	BREMERTON	1.11
	EVERETT	1.13
	KENNEWICK	0.97
	OLYMPIA	1.12
	OTHER	1.06
	PASCO	0.97
	RICHLAND	0.97
	SEATTLE	1.15
	SPOKANE	0.92
	TACOMA	1.14
	VANCOUVER	1.06
	YAKIMA	0.98
WI	APPLETON	1.03
	BELOIT	1.05
	EAU CLAIRE	1.03
	FOND DU LAC	1.05
	GREEN BAY	1.03
	JANESVILLE	1.05
	KENOSHA	1.06
	LA CROSSE	1.03
	MADISON	1.05
	MILWAUKEE	1.08
	NEENAH	1.03
	OSHKOSH	1.03
	OTHER	1.05
	RACINE	1.06
	SHEBOYGAN	1.03
	SUPERIOR	1.05
	WAUKESHA	1.08
	WAUSAU	1.03

State	Metropolitan Area	Multiplier
WV	BECKLEY	0.97
	CHARLESTON	0.98
	CLARKSBURG	0.97
	HUNTINGTON	0.99
	MORGANTOWN	0.97
	OTHER	0.97
	PARKERSBURG	0.98
	WHEELING	0.94
WY	CASPER	0.86
	CHEYENNE	0.86
	LARAMIE	0.86
	OTHER	0.86
	ROCK SPRINGS	0.86
	SHERIDAN	0.86

BNi **Building News**

A **DESIGN COST DATA** COMPANY

DATA YOU CAN TRUST

Home Builders Square Foot Tables

The costs listed in these tables are intended to provide typical, average ranges of costs for the most common types of single family residential construction. Differences in materials and methods of construction will always vary the costs.

Two home types are included: one story and two story. Each is then presented in two quality categories: average, or "contractor grade," and deluxe, or "architectural grade." Low, medium and high ranges of costs are then presented in seven typical square footages.

Each of the examples is further broken down.

The "Frame" includes the foundation, rough framing, siding, roofing, exterior doors and windows. "Interiors" includes all finish carpentry, wall and floor finishes, cabinetry, plumbing, HVAC and electrical. The "Overhead and Profit" includes the markups for insurance, taxes, office overhead and profit for the general contractor.

These costs represent typical averages and do not include land costs, sitework (other than foundation excavation), landscaping or other costs not directly attributable to construction.

ONE STORY AVERAGE

LOW

SQUARE FOOTAGE	1500	1750	2000	2250	2500	2750	3000
COSTS							
FRAME	135.52	132.81	130.10	129.42	126.17	122.94	121.66
INTERIORS	129.63	127.04	124.44	123.79	120.68	117.60	116.37
OVERHEAD & PROFIT	29.46	28.87	28.28	28.14	27.43	26.73	26.45
TOTAL	$294.61	$288.72	$282.82	$281.35	$274.28	$267.27	$264.47

MEDIUM

SQUARE FOOTAGE	1500	1750	2000	2250	2500	2750	3000
COSTS							
FRAME	149.07	146.09	143.11	142.36	138.79	135.24	133.82
INTERIORS	142.59	139.74	136.89	136.17	132.75	129.36	128.00
OVERHEAD & PROFIT	32.41	31.76	31.11	30.95	30.17	29.40	29.09
TOTAL	$324.07	$317.59	$311.11	$309.49	$301.71	$294.00	$290.92

HIGH

SQUARE FOOTAGE	1500	1750	2000	2250	2500	2750	3000
COSTS							
FRAME	155.85	152.73	149.61	148.83	145.09	141.39	139.90
INTERIORS	149.07	146.09	143.11	142.36	138.79	135.24	133.82
OVERHEAD & PROFIT	33.88	33.20	32.52	32.36	31.54	30.74	30.41
TOTAL	$338.80	$332.02	$325.25	$323.55	$315.42	$307.36	$304.14

ONE STORY DELUXE

LOW

SQUARE FOOTAGE	1500	1750	2000	2250	2500	2750	3000
COSTS							
FRAME	169.40	166.01	162.62	161.78	157.71	153.68	152.07
INTERIORS	162.03	158.79	155.55	154.74	150.85	147.00	145.46
OVERHEAD & PROFIT	36.83	36.09	35.35	35.17	34.29	33.41	33.06
TOTAL	$368.26	$360.90	$353.53	$351.69	$342.85	$334.09	$330.59

MEDIUM

SQUARE FOOTAGE	1500	1750	2000	2250	2500	2750	3000
COSTS							
FRAME	186.34	182.61	178.89	177.95	173.48	169.05	167.28
INTERIORS	178.24	174.67	171.11	170.22	165.94	161.70	160.00
OVERHEAD & PROFIT	40.51	39.70	38.89	38.69	37.71	36.75	36.36
TOTAL	$405.09	$396.99	$388.88	$386.86	$377.14	$367.50	$363.65

HIGH

SQUARE FOOTAGE	1500	1750	2000	2250	2500	2750	3000
COSTS							
FRAME	203.28	199.21	195.15	194.13	189.25	184.42	182.48
INTERIORS	194.44	190.55	186.66	185.69	181.03	176.40	174.55
OVERHEAD & PROFIT	44.19	43.31	42.42	42.20	41.14	40.09	39.67
TOTAL	$441.91	$433.08	$424.24	$422.03	$411.42	$400.90	$396.71

TWO STORY AVERAGE

LOW

SQUARE FOOTAGE	1500	1750	2000	2250	2500	2750	3000
COSTS							
FRAME	115.97	113.65	111.33	110.75	107.97	105.21	104.10
INTERIORS	110.93	108.71	106.49	105.93	103.27	100.63	99.58
OVERHEAD & PROFIT	25.21	24.71	24.20	24.08	23.47	22.87	22.63
TOTAL	$252.11	$247.06	$242.02	$240.76	$234.71	$228.71	$226.31

MEDIUM

SQUARE FOOTAGE	1500	1750	2000	2250	2500	2750	3000
COSTS							
FRAME	127.57	125.01	122.46	121.82	118.76	115.73	114.52
INTERIORS	122.02	119.58	117.14	116.53	113.60	110.70	109.54
OVERHEAD & PROFIT	27.73	27.18	26.62	26.48	25.82	25.16	24.89
TOTAL	$277.32	$271.77	$266.22	$264.84	$258.18	$251.58	$248.95

HIGH

SQUARE FOOTAGE	1500	1750	2000	2250	2500	2750	3000
COSTS							
FRAME	133.36	130.70	128.03	127.36	124.16	120.99	119.72
INTERIORS	127.57	125.01	122.46	121.82	118.76	115.73	114.52
OVERHEAD & PROFIT	28.99	28.41	27.83	27.69	26.99	26.30	26.03
TOTAL	$289.92	$284.12	$278.32	$276.87	$269.92	$263.02	$260.26

TWO STORY DELUXE

LOW

SQUARE FOOTAGE	1500	1750	2000	2250	2500	2750	3000
COSTS							
FRAME	144.96	142.06	139.16	138.44	134.96	131.51	130.13
INTERIORS	138.66	135.88	133.11	132.42	129.09	125.79	124.47
OVERHEAD & PROFIT	31.51	30.88	30.25	30.10	29.34	28.59	28.29
TOTAL	$315.13	$308.83	$302.53	$300.95	$293.39	$285.89	$282.89

MEDIUM

SQUARE FOOTAGE	1500	1750	2000	2250	2500	2750	3000
COSTS							
FRAME	159.46	156.27	153.08	152.28	148.45	144.66	143.14
INTERIORS	152.52	149.47	146.42	145.66	142.00	138.37	136.92
OVERHEAD & PROFIT	34.66	33.97	33.28	33.10	32.27	31.45	31.12
TOTAL	$346.64	$339.71	$332.78	$331.05	$322.73	$314.48	$311.18

HIGH

SQUARE FOOTAGE	1500	1750	2000	2250	2500	2750	3000
COSTS							
FRAME	173.95	170.47	166.99	166.12	161.95	157.81	156.16
INTERIORS	166.39	163.06	159.73	158.90	154.91	150.95	149.37
OVERHEAD & PROFIT	37.82	37.06	36.30	36.11	35.21	34.31	33.95
TOTAL	$378.16	$370.59	$363.03	$361.14	$352.06	$343.06	$339.47